BUILDER'S
INSTANT ANSWERS

BUILDER'S
INSTANT ANSWERS

Sidney Levy
R. Dodge Woodson

McGRAW-HILL

New York   Chicago   San Francisco   Lisbon   London
Madrid   Mexico City   Milan   New Delhi   San Juan
Seoul   Singapore   Sydney   Toronto

The McGraw·Hill Companies

Cataioging-in-Publication Data is on file with the Library of Congress.

Copyright © 2004 by The McGraw-Hill Companies, Inc. All rights reserved. Printed in the United States of America. Except as permitted under the United States Copyright Act of 1976, no part of this publication may be reproduced or distributed in any form or by any means, or stored in a data base or retrieval system, without the prior written permission of the publisher.

1 2 3 4 5 6 7 8 9 0   DOC/DOC   0 9 8 7 6 5 4 3

ISBN 0-07-139513-X

The sponsoring editor for this book was Larry S. Hager and the production supervisor was Sherri Souffrance. It was set in Stone Sans by Lone Wolf Enterprises, Ltd.

Printed and bound by RR Donnelley.

McGraw-Hill books are available at special quantity discounts to use as premiums and sales promotions, or for use in corporate training programs. For more information, please write to the Director of Special Sales, McGraw-Hill Professional, Two Penn Plaza, New York, NY 10121-2298. Or contact your local bookstore.

 This book is printed on recycled, acid-free paper containing a minimum of 50% recycled, de-inked fiber.

This book is dedicated to professional builders and their weekend equivalents who need a quick and convenient resource to check on materials or processes necessary to complete their project— whether it be large or small.

Hopefully the reader will find some helpful tips gleaned from the writer's own construction experience to make that job a little easier.

ABOUT THE AUTHORS

SIDNEY LEVY

Sidney Levy has been associated with the general contracting business for 40 years, working on projects ranging from residential high-rises to pharmaceutical manufacturing facilities. He is the author of eleven previous books pertaining to project management, subcontracting, the Japanese construction industry, and international infrastructure. Mr. Levy currently is the principal in a construction consulting business in Baltimore, Maryland where he resides.

R. DODGE WOODSON

R. Dodge Woodson has built has many as 60 houses a year. He is a licensed builder, remodeler, designated real estate broker, master plumber, and master gasfitter. His experience as a land developer and builder makes his contribution to this book extremely valuable. Woodson has authored numerous books and continues to apply his trade as he has for more than 25 years.

CONTENTS

BUILDER'S
INSTANT ANSWERS

PLANNING ON PAPER

Planning on paper is a sure way of making a project more likely to succeed. The benefits of a solid, written plan are numerous. Yet, some developers don't have much patience for paperwork. They think of the planning as being boring and see it as a waste of time. Well, it is not a waste of time. It can be boring, but it doesn't have to be. Many developers find the planning stage to be one of the most enjoyable aspects of land developing. There is some risk that you will find the planning so fascinating that you will spend too much time on it. You have to know when to accept a plan as being as good as it needs to be. Some perfectionists keep tinkering with their plans for too long. It's possible that a developer who is unsure might use planning as an excuse to postpone the commencement of a project.

Many factors must be assessed as a plan is made for a development. Before a project goes into production there will usually be a plan that has been created by experts. But developers need to create their own plans to give the experts a direction to move in. Not all developers do their own planning, and you don't have to. It's possible to hire people to do the planning for you. But doing it yourself can be quite rewarding. Even if you are not comfortable with doing a complete plan on your own, you can complete many of the components within a plan. The more direction that you can offer your experts, the more likely you are to get a project plan that you like.

THE COMPONENTS

The components of a master plan tend to be consistent. However, different types of projects can call for different types of planning elements. Site features are a prime consideration. Both natural features and constraints of land must be considered. Location may be one of the biggest factors to look at. In terms of location, you might be concerned about how close a property is to schools, shopping, and medical care. You will also want to think about comparable properties in the area that you intend to develop.

After working on a plan that deals with existing features, you must create a plan for the development. Some examples of what you might ask yourself if you are building a residential subdivision could include the following:

- What house style will you allow in the development?
- Are you going to have play areas throughout the development for children?
- Should you plan on paved bike paths?
- Will open areas provide recreational opportunities?
- Do you plan to build swimming pools and tennis courts?
- Will you be creating a high-density development?
- Would residents appreciate a community meeting hall within the development?

Good developers ask themselves hundreds of questions during the planning stages.

People who don't have experience in land development generally have no idea how much effort is required to create a working plan. Anyone who has not developed land may not realize that

fastfacts

If you are not an experienced developer, consider hiring a professional consultant to work with you when planning a project. He or she could be an engineer, an experienced developer or builder, or a real estate professional. The benefit gained from the experience of others can be tremendous.

someone has to think of everything from a name for the development to how the streets will be lighted. The amount of thought that goes into a development is tremendous. Minor details can have a major effect on a development. And failure to recognize a need can be extremely costly.

Demographics and psychographics both need to be considered when defining a neighborhood. The age, income, and family status of people can have a major influence on the type of development required. Developers must consider as many factors as possible to determine how to structure a development. For example, recreational needs vary with age. This could affect the need for walking paths or racquetball courts. Knowing your target market is the only way to build a development that can be sold at maximum profit.

SMALL DEVELOPMENTS

Small developments have become popular. Residents often prefer a small development since it preserves a feeling of a close community. Builders sometimes feel that small developments are less expensive. In terms of total cost, a small development does cost less than a large one. However, when the development cost is divided by the number of units available for sale, a larger development will often cost less per unit. This is due to the fixed cost of some expenses that remain essentially the same regardless of the size of the development.

Extremely small developments don't usually require the same types of expenses that larger developments do. Your budget may influence the size of your projects. But it can be just as easy, if not easier, to finance a large development as it is to finance a small one. Tiny developments can fail since there are so few units to be sold. This is a risk with larger developments too, but there may be safety in numbers. By this, I mean that if you have 20 building lots and two that won't sell, you will probably survive holding the two lots. But, if you have four lots to sell and can't sell two of them, you could be in deep trouble.

The appeal of small developments for buyers is the hometown feel. Large developments often come off as cold and impersonal. If you are going to create small developments, you should plan on putting many personal touches into your planning. For example, the entry to the subdivision and the interior streets should be identifiable with the concept of the development.

fastfacts

Small, personal subdivisions can be extremely popular and very profitable. If you decide to go this route, budget enough money to give the subdivision a down-home feel. Putting some special finishing touches on the development can greatly increase demand for it.

YOUR JOB

Your job as a developer requires you to come up with a plan that is unique. It should be a plan that cries out for attention and offers plenty of appeal to your target market. The size and scale of a development must be planned. When establishing size and scale you should consider whether they will be set based on vehicular traffic or foot traffic. Development scale should be in proportion to the proper setting. Don't pick one set of criteria and use it throughout the community. It is better to vary and balance size and scale to keep a development more interesting.

Construction materials can have a lot to do with the tone of a development. What type of look will you be hoping to achieve? Are you going to use a stone wall and stone pillars for the entrance to your development? Would a more modern look suit the nature of the development better? Many types of materials can be used to create different moods. Your development could be rustic or contemporary. It could be based on a high-tech motif or a classical setting. You and your experts have to decide what image you wish to create and then choose materials that will allow you to accomplish your goal.

Land planning is an area for experts to help developers with, but you can start the process on your own. Architectural styles and land planning combine to define a neighborhood. You should strive to create a combination that will not become boring. Yet you should not diversify to a point where the development loses its overall tone. Let's look at an example of how you might create a unique development.

Assume that you are creating a development on the outskirts of a city. Your land is rural, rolling land that offers a country feeling, even though it is close enough to the city for a reasonable commute. The project is fairly large and will combine different types of living opportunities. There will be detached homes, condos, and duplexes. Based on the natural land conditions, the parcel looks perfect to house a development of highly styled farmhouse designs. This design allows for large areas of living space and many options in exterior treatments. The building lots are large enough to accommodate big houses and ample parking.

Your development will be divided into sections for each type of housing to be built. Single-family homes will be in one area, condos will be in another, and duplexes will be in the third section of the development. Since farmhouses are large by tradition, you will be able to build condos and duplexes that will give the appearance of a single-family home. Creative placement of garages and entry doors will make it difficult to tell one type of housing from another when driving through the subdivision. You can mix up the look of the development by using porches, varying roof designs, entryways, attached storage areas, garages, and so forth. All the housing will be in the theme of farmhouses, but there will be enough variation to keep the subdivision desirable.

As a continuation of the farmhouse theme, you might plan to use rustic appointments in common areas. For example, if there is a stream for a footbridge to pass over, you might consider making it a covered bridge. Bicycle racks might be built to look like hitching posts. Water fountains could be encased in small structures that resemble a covered well where a bucket hangs from the rafters. The creative possibilities are usually limited only by financial constraints. You can use census data to help you decide what direction to take.

YOU MAKE THE RULES

As a developer, you make the rules that must be followed by people buying into your development. It is your decision on how covenants and restrictions will be used to maintain the look of your development. Your work should cover acceptable house designs, minimum square-footage requirements, allowable paint and stain colors, the types and colors of roofing materials, and so forth. Some developers go so far as to dictate the types of mailboxes residents may use. Without strong development guidelines, a development can lose its appeal quickly. You, the developer, are responsible for creating and protecting the development. Of course, you will probably turn to

fastfacts

Your decisions on covenants and restrictions in deeds will set the tone for your development. Seek expert guidance when creating these elements. At the very least, work with a good attorney. Consider consulting land planners and architects for concepts to include in the rules for your development.

land planners and architects to assist you in the final stages of developing your guidelines.

LAND USE

Guidelines for land use may begin with setback requirements. The setback refers to the distance that a building must be, at a minimum, from some other object, such as a property boundary. There are front, back, and side setbacks. For example, you may stipulate that houses must be at least 35 feet from any community sidewalk or street curb. You could say that no buildings may be erected closer than 15 feet to any side property line. And you might require a rear setback of 25 feet between a building and the back property line.

Another means of land control is to set a standard for how much of a building lot may be consumed by homes, garages, parking areas, and other impervious surfaces. A rule such as this can guarantee a minimum amount of green space around all homes. Then you move into regulations for site improvements, such as private sidewalks, patios, decks, and so on.

When you establish the rules that builders and homeowners must live by when in your development, you have to be very specific. For example, how wide can a private sidewalk be? Does your development have minimum and maximum widths on record? Wouldn't it look strange to have most sidewalks 3 feet wide and then to encounter some that were 5 feet wide? Will walkways have to be made of concrete, or will gravel walkways be allowed? Consistency is important in development. Are covered entryways going to be allowed in your development? Will they be required? You must decide.

Here are some other topics to consider when you are setting up rules and regulations:

- Porches
- Patios
- Decks
- Swimming pools
- Storage buildings
- RV storage
- Lawn care
- Driveways
- Lighting

Developers can dictate all sorts of rules for their developments. Too many restrictions can scare buyers away. But too few restrictions will make buyers nervous that a development may not maintain its standards. Even the smallest details, such as the type and style of house numbers used to display an address, should be considered when developing your master plan for a project.

Building Regulations

Building regulations are common for subdivisions, and they are needed. Depending upon the type of development, a developer may prohibit the construction of certain types of houses. You might prohibit the construction of single-level homes or homes built on slab foundations. Your rules might prevent builders from constructing houses that don't follow a specific theme. Rules could cover roof pitches, square footage, exterior trim, paint and stain colors, roofing materials, window designs, and so forth. A developer could require that all homes built in a development have brick foundations. The use of vinyl siding might be prohibited in a development.

CONSTRUCTION REGULATIONS

Construction regulations must be planned in advance. Once a project is started, the construction requirements become a means of controlling workers and staying on time and on budget. Some builders and developers tend to overlook the planning of construction rules. Don't fail to allow provisions for construction procedures as you build your master plan. What types of things do you need to be concerned about? Let's find out.

Temporary Roads

You may need to build temporary roads to get workers into a project early. Don't omit this expense from your cost projections. Make arrangements early for allowing workers access to your project. Getting a project ready to begin construction and then realizing that your crews can't get to the job site is not only embarrassing; it sets you back on your schedule, and this cuts into your profits.

Temporary utilities are likely to be needed for your project. This will not always be the case, but determine if any utilities will be needed during the development process. It can take several weeks or more to get temporary electrical service to a building site. If you think that this is a long time, just imagine how long it might take to provide services to a full development. You probably won't run into major needs for temporary utilities unless you are acting as the building contractor and the developer.

Depending on site location and conditions, you may have to install some type of retention system to control erosion. If construction equipment creates a lot of dust in a populated area, you may have to contract with someone to provide dust-control services, such as sprinkling construction roads. All of these types of needs must be accounted for in your master plan.

Site Needs

There are certain site needs that may be required for your project. For example, portable toilet facilities are needed. Noise control is also a potential concern. Trees that are to be saved should be protected with temporary barriers. Some provisions should be made for the storage of equipment and materials. For material storage, many developers use the trailers that are pulled by 18-wheelers. Construction office space is needed, so a site trailer should be placed on the property. Some developers buy their storage and office trailers, while others rent them.

Fenced enclosures help to protect the expensive equipment used when developing land. Not all developers go to this expense, but you might want to consider the option. Provisions must be made to control storm water and erosion during construction and development. A place should be established for the posting of required documents, such as permits and safety posters. Temporary signs will be needed, and you may need someplace to post them. Sit down and figure out all that you can about your needs.

Management

The management of a project is crucial to the success of a development deal. Someone must make sure that all permits are obtained and posted. Keeping active insurance on a project is essential. Any subcontractors used for a project must be screened for insurance and general business compliance. Time schedules must be set. Work hours for contractors must be established. The management needs of a project can be extensive.

You must decide if you will serve as your own project manager or if you will hire one. Good project managers don't work for peanuts. If you are going to hire one, you must factor the overhead cost into your overall budget. Maybe you are not sure if you can do the work yourself. If you have any doubts about your abilities, factor in the cost of a project manager. It's better to have the overhead factored into your budget and not need it than it is to have the expense missing and then find that you need a manager.

The planning stages of a development can be tedious. If you prefer to have it done for you by experts, go that route. Most developers do their preliminary planning on their own, and it's good experience. Once you have a general plan, you can turn it over to the experts. Then, as results come in from the experts, you can compare the final plan with your rough draft. Look to see how much has changed. When you get to the level where changes are minor or extremely technical, you will know that you have a good handle on what it takes to be a successful developer.

Regardless of how your master plan is completed, you will find yourself working with engineers. This can be a daunting experience. However, you should not let the work of engineers go across your desk without some form of review. You might not know what you are looking at when you see it. Ask questions if you don't understand an engineering report. You're paying the bills, which entitles you to know what you are buying

chapter 2

FLOOD ZONES
AND WETLANDS

Flood zones, wetlands, and other deal-stoppers can be a developer's worst nightmare. There are simply some elements of land that can't be worked around. Developers who buy land without adequate research can wind up in deep financial trouble. I guess I have been lucky. While I've come very close to being devastated by land problems, I've never taken a direct hit. Even so, I have lost money that I had not planned on spending due to land elements. Developers who deal with environmental issues have to be able to take a lot of heat and maintain their cool.

Most developers shy away from anything that might be close to an environmental issue, but there are some who have no fear. Then there are those who are willing to walk the line and hope that they don't get too close to the edge. Over the years, I have done my share of line walking, and I've known some developers who have shown total disregard of environmental laws.

Personally, I have a great respect for environmental issues. There is not enough money to make me destroy natural resources that should be protected. When I say that I have worked the edge, I simply mean that I've built in areas where I have taken heat and been proven to be right. The last house I built for myself is an example of such a situation, and it's a good example of how even careful builders and developers can wind up in hot water, so let me tell you a little about it.

I bought 25 acres of land to build my personal home. The 7 acres on which I chose to build fronted on a small river. The river was in view of the home location, but I couldn't see any flood risk or wetlands issues. However, to be safe, I asked the Department of Environmental Protection (DEP) to inspect the site. A representative from the DEP came out and assured me that there was not a threat to the river or the wetlands from my proposed construction. My next step was to have the local building inspector come out for a site visit.

The local building inspector assured me that the land was fine to build on and that getting a building permit would not be a problem, subject to a soils test. I had a soils test done and it was fine. Everything seemed to be okay. Once I was convinced that the deal was safe, I removed all contingencies from my purchase contract and bought the land. A few weeks later, when I applied for my building permit, the permit application was denied. The reason I was given was that the lot was not a legal building lot. I was told that it was an illegal subdivision of a larger parcel. A few weeks ago I had been told that the lot was fine, and all of a sudden, I was in a world of trouble.

Without detailing all of the steps required, I went to the zoning board. The next step was the local board of appeals. My attorney and I had done exhaustive research to prove that the town was wrong in its opinion of the land status. Long story short: I won my appeal and got my building permit. Now I'd done everything reasonable to make sure that the parcel was an approved building lot and still ran into roadblocks. Due to my experience, knowledge, and persistence I won. If I had been an average homebuyer I might have given up.

After the fiasco of the building permit I thought that my troubles were over, and they were, for a while. When I financed the home, I got a construction loan that would convert to a permanent mortgage. The house was finished and the permanent mortgage went into effect. After several months I decided to refinance the loan for a better interest rate. When I did, another problem came up. The bank told me that the house was built in a flood zone. I was shocked and didn't believe it. A survey crew was sent out by the bank.

Then bank's survey crew told me that the house was "probably" in the flood zone. I looked into getting flood insurance and found that since the town where my home was located didn't participate in the flood program, I could not get flood insurance. I was steamed. Then I found out that I couldn't sue the town since they were not insured and that a legal tort protected such towns. Things were going from bad to worse.

I was in jeopardy of not being able to get my new loan. Both the new bank and the old bank were on my case. I hired a surveyor who

fastfacts

Working around wetlands can be very risky for a developer. The smallest amount of vegetation could trigger an environmental issue. If you have any reason at all to be suspicious of a wetland issue, call the proper authorities and have them render an opinion on the property before you buy it or build on it.

just happened to be on the town council to do a full survey of, not only my land, but of the entire river area. It was very expensive, but it proved that my house was not in danger of flooding. This resolved the issue for good. My expense was considerable and the mental anguish was extreme, but I won. All this fighting should have been unnecessary. The land I bought was not a problem piece, but it turned into one.

As you can tell from my personal story, even experienced builders and developers can do all that is reasonable and still wind up in a mess. You might just imagine what could happen to a developer who was careless. Environmental issues are a big factor in land development, so let's get down to the nitty gritty.

WORKING WITH WETLANDS

Working with wetlands is a high-risk venture. Any land containing even a small section of wetlands is a potential time bomb for a developer. I've seen tiny frog ponds kill development deals. In fact, I can remember a piece of land where the mere presence of ferns and cattails scared off a major developer. It doesn't take much to put a piece of property under the scrutiny of environmental concerns.

Several laws pertaining to environmental issues are on the books. There is a law that prohibits unauthorized obstruction or alteration of any navigable water. The law keeps developers from filling in such waters as well as preventing the excavation of material from the waters. This is a serious situation and most developers respect it. Remember when I told you earlier that I knew of developers who had no respect for laws? Well, this particular law was violated by a developer I used to know. The developer filled in wetland areas for development. He did the fill fast and knew that he would be caught.

I had lunch with him after the fact and we discussed his actions. At the time, he was in deep legal trouble. His attitude was that it was his land and that he would do what he wanted to with it. He told me that he had filled the wetland quickly so that the authorities would have to make him remove the fill rather than stop him from placing the fill. The last I heard, the developer was juggling court dates and fighting his fight. Personally, I don't agree with what he did, but he did it.

Another law on the books has to do with the risk of discharging pollutants into navigable waters. Then there is the law that deals with the transportation of dredged materials headed for disposal in an ocean. One of the laws that impacts land developers most is the Emergency Wetlands Resources Act of 1986. This law ensures the conservation of wetland resources. Any wetland area can fall under multiple laws.

Any developer wishing to fill in a wetland area must apply for a permit. The permit application will be reviewed by both the U.S. Army Corps of Engineers (Corps) and the Environmental Protection Agency (EPA). Getting approval for such a fill request is unlikely. To obtain approval, you must demonstrate that you have no practicable alternatives to your filling of the wetlands. Further, you must prove that your fill will not cause significant damage to the aquatic ecosystem. The EPA has veto power over the Corps in such matters.

What is a wetland? You could probably get many answers to this question, but the definition given by the governing bodies is the definition that matters most. According to the environmental authorities a wetland area is an area that is inundated or saturated by surface or groundwater at a frequency and duration sufficient to

fastfacts

Obtaining approval from one agency is not a safe assurance that your building site might not be in violation of some requirement from another agency. Research every project thoroughly to make sure that you are not walking into an ambush.

support, and that under normal circumstances do support, a prevalence of vegetation typically adapted for life in saturated soil conditions. Wetlands generally include swamps, marshes, bogs, and similar areas. What you have just read is the official definition of a wetland, but don't assume that the definition given is all that there is to the matter. The interpretation of the definition can be broad, so you must be cautious.

Wetlands may be regulated by state, federal, and local agencies. Passing muster with one agency doesn't exempt you from the others. If you have any reason to believe that your project might fall into a wetland classification, you need to involve experts to remove any doubts or risks that might jump up in your face. If you violate a wetland regulation, you may be liable for either civil or criminal action against you. Trying to beat or cheat the system simply isn't worth the price you may have to pay.

There are many rules that apply to the disturbance of wetlands. If you have plans for clearing land, dredging areas, or filling sections of your development, you will certainly trigger wetland regulations if the land falls under the wetland protection. Other activities can also put you in harm's way, so just don't do anything in any area where you might be nailed for a wetland infraction.

FLOOD AREAS

Flood areas are bad for developers. The benchmark for flood areas is usually the 100-year flood boundary. This is an area of land that has been flooded within the last 100 years. Most communities participate in a flood program that allows homeowners to acquire flood insurance at reasonable rates. Even at the reasonable rates, flood insurance is an expense that is not required for properties that are not considered to be at risk of flooding. If you create a development where flood insurance is required, selling the lots could be difficult. On the outside chance that you get caught up in a deal where flood insurance is needed but not available, you are in deep trouble.

Local authorities normally have flood maps available for inspection. However, the maps may be old and difficult to read. This is the problem that I ran into with my personal home. The flood maps were old and were not drawn in great detail. It took a detailed survey, which I had to pay for, to change the mind of the powers that be. Have your engineers establish local flood areas and make sure that your building sites are not in them.

HAZARDOUS WASTE

Hazardous waste is a component of modern land developing that old-time developers didn't have to worry so much about. Times have changed, and hazardous waste is a serious consideration in modern development practices. The EPA offers a list of materials that are considered to be hazardous. Materials not listed may also be considered to be hazardous if they exhibit any of the following characteristics:

- Corrosivity
- Reactivity
- Toxicity
- Ignitability

If you become involved with hazardous wastes, you may have a lot of hoops to jump through. The most innocent piece of property can harbor hidden waste and high clean-up costs. If you violate regulations pertaining to hazardous wastes, you might have to foot the bill for all clean-up costs, which will not be cheap. In addition to the clean-up costs you could be hit with huge fines.

There is no way that I know of to be absolutely sure that a parcel of land might not be hiding hazardous waste. Research of past use of the land is about the best defense that you have. It is possible to do expensive scans of the property, but if general research doesn't raise any red flags, there is probably no need for high-tech, expensive scans.

OTHER ENVIRONMENTAL CONCERNS

Other environmental concerns for developers could range from destroying natural habitat for an endangered species to erosion. The list of potential risks can be a long one. This is why you need to bring in an environmental expert to clear your project before you go too far in the developing process. Specialists can be expensive, but they are a real bargain if you compare the risks and costs of what could happen without them.

So far we have talked about hard-line environmental issues. The subjects covered to this point are dealt with under some form of legal protection. But there are other issues that you could face as a developer that are not so clearly defined in legal documents. Sometimes a developer's worst enemy is the public, and you might find yourself in some situation where public disapproval is your biggest drawback to a development.

ANGRY CITIZENS

Angry citizens can be extremely difficult to deal with. Your development plans can meet all requirements, be signed off on, and still fall into a trap that is hard to emerge from. I've had this happen on a small scale, and I've known other developers who have dealt with the problems on a much larger scale. People don't always embrace new developments. While there may be no legal reason or way for the public to stop your development, people can certainly make the profitability of your project suffer. Let me give you a few quick examples from my past that will highlight this facet of developing.

I bought some leftover lots in a subdivision to build on. The subdivision was several years old, and residents had become accustomed to using the vacant lots for their own purposes, such as mulch disposal. When I bought the lots, there was some distress that the vacant lots were about to grow houses. One neighboring resident was especially nasty about my company building in the area. The resident was rude to my workers and would not cooperate with us in any way. When I had surveyors stake the house out so that footings could be dug, the survey stakes came up missing the next day. I paid the surveyors to stake out the house again, and again the stakes were gone when the backhoe operator arrived the next morning. Finally, I paid a third time to have the house staked off and had the surveyors drive iron stakes below the ground level. The next morning I was on the site with my metal detector and found the stakes for the house corners. This problem cost me a few days and a few hundred dollars more for extra surveys, but I finally got the footings in. All in all, compared to some horror stories, my experience wasn't too bad.

I have known of developers who encountered many problems due to unhappy neighbors. The types of problems ranged from people pouring sugar into the fuel tanks of heavy equipment to windows being broken out of houses on a regular basis. If the people near your land don't support your development plans, you could run into expensive vandalism. While I have no first-hand knowledge of anyone forming a blockade or a physical protest on a project, I have heard of such events.

If your project is approved and legal, you have remedies against people who stand in your way, but that might be of little help in the real world. Calling the police daily or filing lawsuits robs you of development time. You may find it prudent to interview residents in the adjacent areas of your development to see if there will be any mass disapproval.

Land development is a business that can be plagued with problems. It is probably impossible to avoid all potential problems. However, with enough knowledge, experience, and research you can dodge most of the bullets. There are plenty of traps waiting for the unsuspecting developer, so keep your guard up and cover all the bases that you can. Some problems are sure to slip through your defenses, but you can head most of them off with proper preparation.

chapter

3

ZONING

Zoning has much to do with what developers are allowed to create. The laws, rules, and regulations for zoning can vary greatly. These variations can run from town to town and city to city. You can't rely on a state-by-state formula when it comes to zoning laws. Every organized community is likely to have its own zoning regulations. And zoning laws can be changed fairly quickly, so don't rely on old zoning decisions. Every piece of land can fall under a different zoning ordinance.

Zoning regulations can be very complicated. Even experienced real-estate attorneys can have trouble interpreting the laws. Trying to make sense on your own of all the zoning issues that you will deal with as a developer or builder would be crazy. You will need a good lawyer to work with you to cut through all the red tape. However, there is a lot about zoning that you can understand and work with. These issues are what we will cover here. But remember to consult your attorney for clear interpretations of zoning ordinances before you make a buying decision on a parcel of land.

ZONING MAPS

Zoning maps are good starting points. One purpose of zoning is to prevent conflicts in land use. It is the zoning laws that balance a community in what is believed to be the best means possible. Zoning maps are drawn to show specific zoning regions. For example, you might see that one part of town is zoned for retail use, while another section is zoned for industrial use and another section is zoned for residential use. It is common for zoning maps to be altered periodically. Requests for zoning changes are sometimes approved. When they are, some reference to the changes must be recorded. Eventually, the zoning maps reflect the changes.

If you are researching a particular piece of land, you can look on a zoning map and see what the existing zoning is. When the established zoning is compatible with the type of project that you want to build, you have it made. It is a simple matter to move forward when you don't need a zoning change to do it. However, if you find that the land that you are considering purchasing is not zoned for a use that will allow your type of development, you have some trouble.

If a change in zoning will be required for a project, you could be looking at many months of legal maneuvering and a lot of out-of-pocket cash expenses. Your choices are to pass on the property and look for another parcel that will be zoned for your needs or to file for a change in zoning. Some types of zoning changes can be reasonably simple to obtain, but others are very difficult. It is certainly easier and less expensive to develop a piece of land that is already zoned for the proposed use.

I have had my share of dealings with zoning officials and zoning boards of appeals. The experiences have not been fun. Fortunately, I can't remember a time when I did not prevail. But this is not to say that winning was easy or that it came without a fight. Some of my zoning battles have been very expensive. They have also been quite time-consuming. If a project does not have enough potential, fighting the system is not worthwhile. Before you engage in a major zoning battle, make sure that the time and money that you will invest in it will be rewarded in the end.

fastfacts

Never begin construction until you are satisfied that the type of structure that you will be building is in full compliance with zoning laws and regulations. Always confirm the required setbacks before digging a foundation.

fastfacts

Builders who dabble in land development can increase their income substantially. There is risk in land development. It is not a business for people who are uninformed or in a hurry to see results. Successful developers settle in for the long haul and often double their money on projects. This can really add to what a builder earns.

Zoning boards usually have a lot of latitude in how they handle their regulations. A landowner might obtain a variance with relative ease. When a variance is approved, it is usually in the form of a minor variation from existing zoning requirements. As an example, a variance might be issued for a garage to be built 2 feet closer to a property sideline than what present zoning requirements call for. It would be common in such a case for a developer to be required to talk with neighbors who might be affected by the variance and to gain permission from the neighbors before the variance is issued.

Zoning maps show you what existing conditions are. The maps do not indicate that the land might fall within an area that allows a different kind of use. In some cases, the zoning maps might tip you off to great opportunities. Finding land where zoning has been changed for a higher and better use can result in much higher profits for a developer. If you can buy land that the owner believes is zoned as residential use only and then use it for commercial purposes, the value of the land should soar. This is not all that uncommon.

Many residential areas slowly change over to other types of land use. This usually happens as an area grows. Car dealers, fast-food restaurants, hardware stores, and all sorts of other non-residential uses move into an area. As this happens, the houses are sometimes converted to new uses, or they may be removed to allow a higher use of the land. People who live in the houses sometimes sell their property for huge profits. But some sellers are not aware of how much more their homes are worth as the zoning laws and land use have changed. Astute developers scour zoning changes in search of rare opportunities. Zoning maps may not show the changes soon enough. Dig deeply when you are researching permitted land use. Your research can keep you out of trouble and may make you much more wealthy.

CUMULATIVE ZONING

Cumulative zoning is a type of zoning that often allows for changes in zoning regulations. Developers tend to like this type of zoning, since it can allow a great deal of freedom. Communities, however, sometimes suffer from cumulative zoning. The purpose of zoning is to manage land use and to maintain certain separations. Some strange results can occur with cumulative zoning. For example, you might find housing developments mixed in with commercial projects.

The value of land is often in direct relation to the zoning laws. Obviously, a piece of land that can have a shopping mall built on it should be worth more money than the same piece of land if it is limited to single-family housing. Striving for the highest and best use is a good goal, but it is one that some communities cannot afford to enforce. The result is often a mixed-use community, which allows different types of land use in the same area.

Why would a community vote for mixed use? Deriving tax dollars from landowners might be one reason. In some areas, strict zoning keeps residential areas so far from places of employment that traffic becomes a problem. If people have to move into fringe areas and commute to work, the traffic flow can be more than the roads in a community can handle. There can be any number of reasons for cumulative zoning.

FLOATING ZONES

Floating zones are usually districts that are not mapped for specific zoning uses. Communities may use floating zones to give themselves flexibility in applying their zoning regulations. Since floating zones are often unmapped, they can be difficult to pin down during preliminary research. A floating zone may be used as a means to move into transitional zoning. Planned unit developments often come into floating zones.

TRANSITIONAL ZONING

What is transitional zoning? It is a type of zoning that starts with one region of land use and gradually changes the use as distance increases. This type of zoning usually works well. It can facilitate a number of land uses without hurting the look of a city. For example, the strictest end of the zoning might contain heavy commercial properties. As a development sprawls out, the zoning could go to light commercial, then to office space, and eventually to residential use. You might even find regulations that require office space to resemble residential architecture.

There is a town in Maine where low-impact features are required of many types of business property. For example, a major fast-food chain is housed in what could appear to be a large farmhouse. Many of the restaurants and businesses are required to be housed in buildings that are compatible with the residential area. This can make it confusing for people who are looking for the business image that they have come to be so familiar with, but it does enhance the quaint appearance of the town.

PLANNED UNIT DEVELOPMENTS

Planned unit developments (PUDs) are quite common. This type of development may house everything from single-family homes to commercial retail stores. It is common for a PUD to be a stand-alone community. The community may include all forms of service businesses, medical facilities of some sort, and a variety of housing types. Due to the mixed use in a PUD, careful planning is needed on the developer's part.

Floating zones are common areas for PUDs. Many communities encourage the development of PUDs. But there is no guarantee of having a project approved, and this is a risk. The money spent designing a PUD for application approval can run into thousands of dollars. This is money that will be wasted if the project is not approved. Gaining approval for a PUD can mean major profits for developers, but the venture capital invested to get the approval can prove to be a substantial risk.

fastfacts

Builders can inch into land developing with small projects. Buying some rural residential land and cutting it into house lots is not a bad way to get started as a developer.

When a PUD is designed, it can contain many types of buildings and land uses. Most commercial uses are required to be neighborhood-serving establishments. Generally, PUDs are required to provide open space, recreational facilities, and other amenities. It is common for some provision to be made for local fire prevention and protection. Schools may also be a part of a PUD. The expense of developing a PUD runs high. But a PUD that is well planned should sell well. Few first-time developers start with PUDs.

When a PUD is designed, sections of the land are set aside for different types of land uses. These sections are identified on a site plan. For example, one section may be given a rating of R-1, which could mean low-density use. A section with a rating of R-20 could be set aside for ultra-high-density, and there could be other ratings between the two extremes.

CLUSTER HOUSING

Cluster housing is popular in large urban areas. The general concept behind cluster zoning is a compromise: smaller house lots are side by side with larger open areas to balance the mix. Some communities approve cluster developments in an attempt to preserve land around historic sites or even valuable farmland. Developers who can gain cluster approval can turn a small parcel of land into a valuable housing project. There usually is not as much flexibility with cluster zoning as there is with planned unit developments.

Zoning density is usually maintained in a cluster development. However, the lot size and setback requirements are often forgiven. There is not as much freedom with a cluster development as you might think. Even though you can reduce lot size, you may not be able to increase the number of housing units. Remember, the goal of clusters is to provide more open space per dwelling unit. Many developers think of clusters as a way of squeezing a lot of housing into a small space. This is sometimes possible, but don't expect it to be the rule. Rather, it is the exception in terms of density.

EXCLUSIONARY ZONING

Exclusionary zoning is intended to prohibit specific types of development. This type of zoning is used to keep a certain type of development out of a region where general zoning would allow that kind of use. Take, for example, a region where clubs are allowed. There might be exclusionary zoning to prevent the creation of a gun club that includes a shooting range. The noise from the firing range might be the reason for excluding that type of land use. Noise, odor, and pollution are common reasons why exclusionary zoning is used.

Gun clubs might not be the only target of exclusionary zoning. This type of zoning can limit any type of land use, at least in theory. For example, the zoning might exclude the construction of apartments or mobile homes. Minimum house sizes might be required. This is not uncommon for a developer to do in covenants and restrictions, but it is odd to have a zoning regulation of this type. In fact, exclusionary zoning often comes under fire as being a form of discrimination.

INCLUSIONARY ZONING

Inclusionary zoning is designed to promote the development of both low- and moderate-income housing. To do this, communities offer more flexibility in their zoning regulations. For example, a developer may be allowed to increase the density of a development in exchange for controlled-price dwellings. This can sound good on paper, but it may not work so well in reality. Not all land parcels can be maximized with the type of housing that you may want. For example, you may have to include townhouse designs to achieve the higher density.

OTHER TYPES OF ZONING

There are other types of zoning. In fact, you may run into any number of zoning situations as you move from one jurisdiction to another. Zoning laws are similar to building and plumbing codes. They are usually offered in a generic form that is adopted and adapted by local governing bodies. The adaptation can be extreme. You and your attorney will have to read all the fine print to stay out of trouble. Don't make any assumptions. Zoning is a major part of the development process, and it is a topic that you will have to become acquainted with.

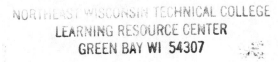

4

SITE EVALUATION

A piece of land can give many different impressions to various people. If four people look at the same piece of property they may see four different images for development. One person might see a golf course while another developer might see a prime site for a shopping mall. A third investor could conjure up an idea for a townhouse development while the fourth person might see a wonderful opportunity for an office park. Some parcels of land cry out for a particular treatment. Others are more versatile and stimulate creative minds. Every piece of property is unique in some way. Successful developers learn to find the highest and best use of land.

Some developers depend on land planners to evaluate land. This is not a bad idea, but many developers prefer to be more personally involved. Most builders like to pick their own parcels for development. Learning to see various aspects of land is a key factor in making wise purchases. Even if your perception of a parcel of land is only the first step in a buying decision, it is extremely helpful to have an ability to spot the differences between prime properties and problem properties.

Average people don't have the education and experience needed to make a full assessment of land. This is not to say that you can't make a reasonable assessment of land that you may wish to develop. Going on a gut reaction is dangerous as a developer, but it is often the first impression that a developer has that proves to be accurate. How do you feel about your ability to look at a piece of raw land and to see it as a developed project? Are you nervous? If you don't have confidence in yourself, you should hire a professional to do your assessments for you. Don't be concerned. As you work in the business you will gain a better understanding of what to look for. If you have strong feelings and a sense of seeing into the future, you are better prepared than many developers.

ROCK

A major consideration to developers is the presence and depth of bedrock. This type of rock is solid and can have a tremendous impact on the cost of development. In some regions bedrock is referred to as ledge. Living in Maine, I run into a lot of ledge when seeking buildable land. Bedrock runs deeper in some regions than it does in others. Many places can be developed without ever having to deal with bedrock. But, some parcels of land have exposed bedrock. If the bedrock is exposed, you should see it when you walk a piece of land. Looking at the lay of the land can indicate the presence of bedrock close to the surface.

Even if you can't see bedrock, you can test for it with simple tools. One way of testing for the depth of bedrock is to drive a pointed steel rod into the ground. If the rod hits rock frequently in different locations there is a good chance that it is hitting bedrock. I've gone to many sites with steel rods, a stepladder, and a sledgehammer to test for ledge. This method can be time consuming and it is not pure proof of what's below the soil, but it's not a bad way to get a good read on a piece of land.

Another way to explore bedrock is to use a backhoe to dig test holes. This is something that should be addressed as a contingency

fastfacts

Do not attempt to dig, probe, or drive rods in ground that may contain buried utilities. Most communities have a phone number that contractors can call to see if a region is safe to dig in. Failure to call such a number that results in damaging underground utilities can be very costly. Call before you explore beneath the land.

fastfacts

Fractured bedrock made of limestone can collect water and col-
lapse downward. If this happens under a road, sinkholes develop.
Do your homework before you build to avoid financial disasters.

clause in a purchase contract. By having a backhoe dig holes in ran-
dom locations you can determine if bedrock will be a problem for
you. Geology maps can also be used to predict what the bedrock
conditions are on a project site. These maps are available from the
U.S. Geological Survey. Local county engineering offices often have
these maps available for your review on the premises.

The problems that may arise from bedrock can be numerous. For
example, if you wanted to build homes with basements and found
that there was bedrock two feet below the ground's surface, you
would have a serious problem. You would either have to scrap your
plans for basements or do a lot of blasting, which would be cost prohibitive. Burying sewer and water pipes can be a big problem if
bedrock is too close to the earth's surface. Roads built over solid
bedrock can be less expensive to build, since compaction is not
needed at much as it would be without the bedrock. However, fractured bedrock made of limestone can collect water and collapse
downward. If this happens under a road, sinkholes develop. This
could also happen under a footing for a building. Before you do a
deal, you will need an engineer to evaluate your site and create a
plan for it, but if you find things on your own that blow a deal out
of the water, you save yourself the cost of an engineer.

SOIL CONDITIONS

Soil conditions can have a strong influence on the cost of a project.
For example, trying to build in sandy soil that runs deep can get
expensive. Footings for construction must be taken down to solid
ground, and this can make for some very expensive foundation
walls. If your development plans call for private sewage disposal, the
perk rate of soil in the septic area will be critical. Not all soils will perk
well enough to permit the installation of a septic field. While there

are ways of working around most types of problems with soils, the cost of doing so can be very prohibitive.

The U.S. Department of Agriculture Soil Conservation Service and the U.S. Geological Survey for your region can be of great help in determining soil types and conditions. Digging test holes is a way for you to do your own on-site soils test. Augers can be used to create deep holes with small diameters. You will need to see the layers of the soil. A shallow test pit might show a sandy soil that would indicate good drainage. Digging a deeper pit may reveal that there is a layer of clay below the sandy soil. The clay would slow the absorption rate considerably. Without an extended knowledge of soils, you will not be able to avoid having a soils expert on your development team for larger properties. But, you can do some preliminary tests yourself before calling in the expert and this could save you some money.

DRAINAGE CONDITIONS

Drainage conditions are always a factor in land developing. Some drainage features can be seen easily. Others are not so easy to spot and require the use of elevation maps for evaluation. Developers can, and do, alter elevations. Dirt is hauled in to some sites can cut out of other sites. But, finding a site that requires a minimum amount of dirt movement is advantageous, both in time and money. When you walk a piece of land you can get a good feel for the drainage issues that may come into play.

If you are considering the purchase of riverfront property for development, you must determine if the construction area will be in a flood plain of flood zone. This may seem like a simple task, but don't be lulled into a false sense of security. Sites that appear to be well away and above a flood risk may not be rated as being out of the danger zone. Local county engineering offices usually have flood maps that you can review. Take advantage of this research opportunity. Buying land that winds up being in a floodplain can be a very costly mistake.

When you are walking land, look for gullies that might fill with water during hard rains, hills that might erode, and depressions that may hold water for long periods of time. Think as you walk and look over the land. Pay attention to what you are seeing. Making sketches if it will help you to maintain a focused view of the lay of the land. Picture where your buildings will be, where your roads will run, and where other elements may be located. Don't hesitate to walk the land more than once. Go over the land time and time again, until you are sure of your feelings about it.

HAZARDOUS MATERIALS

Hazardous materials that are discovered on a piece of land can create a major expense when it comes to clean-up costs. It can be difficult to tell just by looking that a piece of land is affected by hazardous materials. However, the signs of trouble can sometimes be seen if you look hard enough. It would probably be obvious if you saw dozens of 55-gallon drums strewed around a piece of land that you might be in for some trouble. Of course, it would depend upon what was in the drums.

Dump sites are not uncommon on large parcels of land. And remember, small pieces of land started out as large parcels. When raw land is used as an open dump, any number of substances might be discarded on the land. Finding a dump site doesn't mean that you should run in the opposite direction. The site might be mostly harmless, but any dump site is reason for concern.

Underground storage tanks can be expensive to remove. These tanks could be found nearly anywhere. An apartment building might have had underground oil tanks for many years and then converted to aboveground tanks. If the underground tanks are still buried, they are a risk. I've run into this type of situation before. Fortunately, I was able to spot enough evidence to create suspicion before the building was purchased. What tipped me off? When I inspected the basement walls of the building I saw some patch marks that were about the right size to hide the removal of oil pipes.

When I looked in the basement there were oil tanks in it. A quick inspection would not have revealed the evidence of what looked like

fastfacts

Underground fuel tanks have been used extensively over the years. They represent a potentially expensive problem to remedy. Having an environmental scan done on prospective building sites is a good idea when trying to protect yourself from unseen problems.

old piping that had been removed. I followed up on my suspicions by going over the lawn with a metal detector. I found a large metal object with the detector. Then I did some probing with a pointed steel rod. Sure enough, there were two buried oil tanks in the ground. Since I found them early, the seller had to pay for their removal. If I had purchased the property and found the tank later, I'd have been paying the big bucks for the tank removal work.

Farmers may have underground tanks that store gasoline or diesel fuel. If you buy a farm to develop and then stumble upon the underground tanks, you have an added expense, and it can be quite costly. Modern regulations are fairly tight for anyone having underground storage tanks, but the regulations were not always so strict. Older properties can pose quite a risk of underground tanks and other types of hazardous materials. Land that looks perfectly normal could be an expensive nightmare for any developer who buys it. Have all land screened for environmental risks prior to a full commitment to purchasing the land.

LOOKING TO THE PAST

Looking to the past is wise when you are researching a piece of land. Can you imagine buying a large tract of land for development and finding out that it was protected from development due to historical value? Well, it could happen. Make sure that any land that you are planning to develop will not be frozen with red tape. Don't think that you are safe just because other developments have been completed nearby. Each piece of land, and even portions of a piece of land, can fall into a protected status. Getting into a legal battle over the right to develop a piece of land that appears to be protected is not something that many developers would ever wish to do.

OLD WELLS

If you are not familiar with country property, you might not think to keep an eye out for open, abandoned water wells. Falling into an old well could be fatal. I've run across dozens of abandoned wells over the years. Sometimes they have a rotted cover on them, and sometimes there is no cover. One wrong step could send you some 30 feet down into cold water, with little chance of escape. If you see evidence of an old foundation, turn on your well radar—there's probably a well close by. If you can't see clearly where you will be placing your feet, use a staff to probe the ground before you walk. It's wise to take someone with you when you are walking land, and this is even more essential if you suspect risks, such as water wells.

OTHER THINGS

There are many other things that you should pay attention to when you are going over a piece of land. Perhaps, one of the most important is the location of the property. Location is always a prime consideration when dealing with real estate. The size of a parcel is also of interest. Of course, location, size, and price are all issues that you will probably have considered prior to walking a piece of property.

When you walk a piece of land you should study the vegetation. If you have a good understanding of vegetation, the plants growing on a piece of land can speak volumes to you. It's easy to see if trees are on a piece of land. But, smaller vegetation can be very informative. For example, ferns indicate moisture. An abundance of ferns could indicate wetlands or at the least, land that will not drain well. It's not unusual for developers to study books on vegetation to learn what the various types of growth can tell them about a piece of land.

Access to land is not always what it seems. You must make sure that the parcel of land you are considering has suitable access. Finding this out will take some paper research, but you should note site conditions to refer to when you do your deed research. While you are looking around, take notes about the existence of utilities. Do you see fire hydrants? How far away are electrical and telephone wires? Do you see any sewer manhole covers in the road in front of the property? Do surrounding properties have water wells in plain view? Make mental and written notes regarding utilities. These notes will be appreciated when you, or your experts, begin to look

fastfacts

Access to land is not always what it seems. You must make sure that the parcel of land you are considering has suitable access.

into the feasibility and cost of getting utilities to the property for your development plans.

Carry a camera with you on site inspections and take lots of pictures. Photograph surrounding properties, as well as the subject property. The photos can prove very helpful when you research the property values of surrounding parcels. When you leave the land, note the milage on your odometer. You should pay attention to how far the land you are looking at is from various amenities, such as hospitals, schools, and stores.

There is much more to walking a piece of property than just moving across it. A camera and a notebook should be considered required equipment for all site visits.

chapter 5

GROUND CONDITIONS

G round conditions are always a consideration when preparing to build a structure. Deciphering the dirt of a project is a job that is best left to soil scientists and engineers. But it never hurts to have a good idea of what the experts will be talking about. On some sites, the primary concern for soils is that they will perk well enough to allow the installation of a septic system. Large sites can be much more involved. In many cases, the soils of a site control much of the development potential—not to mention the cost of a project. Many factors influence the viability of a project, and the soil conditions are certainly a major consideration.

Most builders and developers are not overly concerned with scientific data pertaining to soils. If the ground will bear foundations, drain properly, and support a successful development, that is generally all that builders and developers are interested in. How much do you want to know about soils? Do you really care that the official name for a soil scientist is agronomist? Probably not. Even if you don't care to know the difference between soil types and characteristics, you should invest some time in understanding the basic terms that you will be dealing with as a developer.

What is soil? Do you know what soil is made up of? There are three components of soil. The first element of soil is the particles that most people think of as dirt. But mixed in with the particles are water and gases. Soil that is fully saturated has all the voids between the particles filled with water. If the soil dries out completely, the voids are filled only with gases. Under normal conditions, a combination of

TABLE 5.1 Swelling and shrinkage of soils.

Type of soil	Percent of swell	Percent of shrinkage
Sand	14–16	12–14
Gravel	14-16	12–14
Loam	20	17
Common earth	25	20
Dense clay	33	25
Solid rock	50–75	0

water and gases fills the voids between the particles. Engineers must assess the phases of soil to arrive at their engineering data. The relationships of gases, water, and particles must be established in a weight-to-volume ratio. Once engineers have the basic evaluation of the soil, they can establish the soil properties for such elements as shear strength, shrinking, swelling, consolidation, and so forth.

The texture of soil is often talked about. Many people refer to soil based on its texture. How many times have you heard people talk about sandy soil or clayey soil? The texture of soil is determined by the relative amounts of the individual particles making up soil. Different tools are used to identify particle sizes. Sieves are used to sort particles by size. Hydrometers are used to measure the amount of soil in suspension. The process is much more complicated than most builders or developers are willing to get involved in. But, hey, that's why we have engineers. Without getting into scientific formulas and procedures outside the realm of duties normally associated with land development, let's just look at how the soils on a project will affect your business.

fastfacts

Soil texture has a lot to do with construction methods. Building on sandy soil will require different tactics than building on clayey soil. Texture is determined by the relative amounts of individual particles contained in the soil.

TYPES OF SOILS

There are many types of soils to be encountered. Many soils are transported in one way or another. The physical aspects of soil types can be judged to some extent by the means of transportation. We're not talking about trucking dirt into a site. No, the transportation referred to here has to do with natural movement.

Alluvial and lacustrine soils are created by sedimentation. Alluvial soils are left by running water. Lacustrine soil is the result of deposits in lakes. Since alluvial deposits are transported by running water, a natural filtering process takes place. Large particles tend to sink to the bottom of a stream, while smaller particles are moved with the water. The separation process is natural and effective. Both alluvial and lacustrine soils tend to make poor foundation materials. Due to their make-up they are generally either medium to fine sands, silty, or consist of clay. The drainage factor for this type of soil is poor. Building a foundation on alluvial or lacustrine soil is risky, since the soils are usually soft, loose, and highly compressible.

Glacial soils are called moraines. The soils are pushed, eroded, or carried along with glacier movement. Particle sizes range from a clay consistency to large rocks. Characteristics of glacial soils can vary greatly. However, glacial deposits generally offer a good base for foundation construction.

Eolian soils are transported by wind. As you might imagine, this means that the soil must be small and light. Sand is the most common type of eolian soil. If you have ever been to a desert or a large

TABLE 5.2 Soil properties.

Soil type	Drainage rating	Frost heave potential	Expansion potential
Bedrock	Poor	Low	Low
Well-graded gravels	Good	Low	Low
Poorly graded gravels	Good	Low	Low
Well-graded sand	Good	Low	Low
Poorly graded sand	Good	Low	Low
Silty gravel	Good	Moderate	Low
Silty sand	Good	Moderate	Low
Clayey gravels	Moderate	Moderate	Low
Clayey sands	Moderate	Moderate	Low

fastfacts

While buildable, the following soils are not great for foundations:
- *Alluvial*
- *Lacustrine*
- *Eolian*
- *Colluvial*
- *Organic*

beach, you have probably seen sand dunes, which are representative of eolian soils. Sand is not a great foundation soil to work with.

Colluvial soils are transported by gravity. In most cases, this type of soil is the result of hillsides deteriorating. Rock chips comprise most of the colluvial soil. Since the soil has moved once, it is likely to move again. This makes colluvial soil undesirable as a foundation material. Organic soil is no better. This is soil that is made up of decaying plant life. Peat is the best known organic soil. Due to the compressible nature of peat, it is unsuitable for foundation construction.

BEARING CAPACITY

The bearing capacity of soil is a primary concern when planning a development deal. You must know what the soil's ability is to support structure loads created during your development. This means buildings, roads, and other improvements. A first concern is the strength of existing soil, but you will also want to know how excavated soil that might be reused on your project will hold up. When embankments will be created, you have to know how stable the soil used to build the embankment will be. If you find that some of the soil is substandard in strength, your engineers might be able to recommend a way of improving its bearing capacity by adding chemicals or materials. In any event, you must establish that the soil will be suitable for your development plans.

Bearing capacity is increased when soil is compacted. When soil is compacted, the void ratio is decreased, which increases the soil's strength. Compaction does many things. Soil that is com-

TABLE 5.3 General bearing capacities of foundation soils.

Material	Support capability
Hard rock	80 tons
Loose rock	20 tons
Hardpan	10 tons
Gravel	6 tons
Coarse dry sand	3 tons
Hard clay	4 tons
Fine dry sand	3 tons
Mixed sand and clay	2 tons
Wet sand	2 tons
Firm clay	2 tons
Soft clay	1 ton

pacted will be stronger and will not be as likely to settle over time. In the case of foundations, settling soil can result in cracks. So, compacted soil that will not be prone to settling is better than soil that is not compressed. Soil used on embankments will be more stable if it is compacted.

To most construction workers, compacting soil is simply a matter of pressing dirt together with either a tamper or a roller. In the most simple of terms, this is true. However, the compaction of soil increases the soil density by rearranging soil particles. Engineers may talk about the fracture of grains of soil and the bonds between them. You might hear the experts talk about bending the soil. Terms like cohesive resistance may be used. What does it all mean? It means compacting the dirt with a tamper or a roller. Sometimes

fastfacts

Why is soil compacted?
- *To increase bearing capacity*
- *To make the soil more stable*
- *To prevent settlement of the soil after construction*

water is added to the soil to improve the compaction rate. As far as most developers are concerned, knowing that compacted soil is stronger and more stable than loose soil is enough to move on with.

Soil requires a moisture content to be compacted to its maximum density. When fill dirt is hauled in, it is usually compacted to accept weight loads. Arriving at maximum density is desirable. The moisture content required for maximum strength varies from soil type to soil type. For example, sandy soil does well with a moisture content of about 8%. Clay, on the other hand, compacts best with a moisture content of about 20%. Project engineers will evaluate fill areas and call for certain compaction specifications as needed to meet the requirements of your development. Maintaining safe loads of soils is essential.

As most builders know, soil compacts best when the soil depth is kept shallow. For example, a plumber would not backfill a sewer

TABLE 5.4   Safe loads by soil types.

Tons/sq. ft. of footing	Type of soil
1	Soft Clay
	Sandy loam
	Firm clay/sand
	Loose fine sand
2	Hard clay
	Compact fine sand
3	Sand/gravel
	Loose coarse sand
4	Compact coarse sand
	Loose gravel
6	Gravel
	Compact sand/gravel
8	Soft rock
10	Very compact
	Gravel and sand
15	Hard pan
	Hard shale
	Sandstone
25	Medium hard rock
40	Sound hard rock
100	Bedrock
	Granite
	Gneiss

fastfacts

*Soil that is being added to an area and compacted should be
added a little at a time. Most soils should be added in layers that
are not more than 8 inches [metric?] thick. Each layer is compacted before the next layer is added. In some cases, water is
applied to the layers of fill to increase the compaction rate.*

ditch with 3 feet of dirt and then run a tamper over it. The ditch
would be filled with layers of dirt, and each layer would be tamped
before the next layer was introduced into the ditch. Compaction in
stages is the best way to arrive at maximum strength.

Soil that is being added to an area and compacted should be
added a little at a time. Most soils should be added in layers that are
not more than 8 inches thick. Each layer is compacted before the
next layer is added. In some cases, water is applied to the layers of
fill to increase the compaction rate. The layers are often called lifts.
Some types of fill, such as gravel and sand, might be added in lifts
that run up to a foot in depth. Compacting soil in layers takes time.
It might seem tempting to fill an area with dirt as fast as possible and
move on with other parts of the development. This would be a
costly mistake. Soil that is not compacted properly is likely to cause
any number of problems for a developer. Don't cut corners on soil
compaction. Factor in the time and cost required to do the job
properly and stick to a proven plan for suitable compaction.

COMPACTION EQUIPMENT OPTIONS

Compaction equipment options exist for developers. What is the
best type of compaction equipment to use on your project? Your
site contractor will be the person who is most likely to decide on the
type of equipment to be used. However, your engineers may specify the type of equipment that must be used to create a satisfactory
compaction. As a developer, you should have a general idea of what
type of equipment is used for various jobs. Rollers are the most common type of equipment used for large-scale soil compaction. But
there are different types of rollers. Should your project be prepared

with a smooth roller, a Sheepsfoot roller, or a vibratory roller? You may decide to leave decisions pertaining to equipment up to your engineers and contractors, but let's take just a few moments to look at the roles of different types of rollers.

Sheepsfoot Rollers

Sheepsfoot rollers use drum wheels that have a large number of bumps or protrusions on them. The protrusions direct a lot of pressure in small areas for tight compaction. Water can be used to fill the roller drum for additional weight. Clay is compacted very well with a Sheepsfoot roller. The bumps on the roller wheel can usually deliver up to 1,000 pounds of pressure per square inch (psi).

Smooth Rollers

Smooth drum rollers are used most often for finish work. Since the roller drum is smooth, it maintains full contact with the soil at all times. Any type of soil except rocky soil can be compacted with a smooth roller. Since the roller drum maintains full contact, it does not create a tremendous about of pressure per square inch. Compare a Sheepsfoot roller at 1,000 psi compaction to a smooth roller at about 55 psi.

Rubber Rollers

Rollers with rubber tires might not seem like much of a compaction tool, but they are. Pneumatic tires are spaced close together with a rubber roller. Weight is added to the equipment to obtain compaction pressures up to about 150 psi. You can get up to 80% coverage with a rubber-tired roller. However, this type of roller should not be used for initial compaction of some clay soils.

Vibrating Rollers

Vibrating rollers are often used in roadwork. Like other rollers, vibrating rollers compress soil with the weight of the roller, but as an added bonus, the pounding of the vibrating roller packs soil even more tightly. Both granular soil and rock fill can be compacted with a vibrating roller. The roller must be set up properly for the type of material being compacted. For example, rock fill would require the roller to be set for a heavy weight with a low-frequency vibration. Sand, on the other hand, would call for the roller to produce light to medium weight and high-frequency vibrations.

Power Tampers

Power tampers are used where compaction areas are small and difficult to gain access to. The tamper is usually gas-powered and operated by one person who walks behind the equipment. When trenches are backfilled, power tampers are most likely to be used. Since the tampers are small and fairly easy to handle, they can be put in a small trench and operated by a single person. Any type of inorganic soil can be compacted with a power tamper.

STABILITY

The stability of soil is directly related to its shear strength. What is shear strength? It is a rating of soil that is determined by calculating the resistance to sliding between soil particles. Physical characteristics of soil determine the shear strength potential. For example, the confining pressure, surface roughness of particles, and soil density all affect the shear strength of a soil. Voids between soil particles are known as pore spaces. Water can flow through pore spaces and affect shear strength.

The stability of slopes must be considered carefully. Embankments fail for various reasons. The strength of any fill material on a slope can affect slope stability. Existing soil strength must also be considered in terms of slope failure. Drainage on a slope is another factor to consider when evaluating the possible failure of an embankment. Engineers can devise methods for stabilizing most types of soil. Building an embankment is not as simple as just piling up dirt. Erosion and general failure must be assessed.

TAKEN FOR GRANTED

The consistency of land is often taken for granted by builders and developers. People walk over a piece of land and don't often give much thought to the soil under their feet. Once development is started, the soil issues can become much more intense. A lot of money can be lost if the reading of the soil for a project is not done correctly. Developers cannot afford to skim over soil issues. Someone has to dig deeply into all aspects of soil characteristics. Search out qualified experts and let them to their jobs. Don't attempt to become your own expert in soil analysis, but learn enough about soils so that you can understand most of what your experts present to you.

6

ENGINEERING EVALUATIONS

Engineering evaluations are a part of any major development deal. Most developers consider the reports as a needed part of the process but are not interested in reading or understanding them. It's common for developers to want a report summary that is easy to read and understand. For most developers, a summary is enough. There are, however, some developers who prefer a much more detailed report to study. Many developers hardly do more than scan their engineering reports before they are passed on or filed. Will you be a developer who learns to understand the reports or one who will simply take the advice offered in a summary report?

Working with summary reports is faster than digging through piles of technical terms. But there is something lost in the translation, so to speak. Abridged versions of reports obviously omit certain details. If you are the type of person who wants to know every aspect of your developing process, you will want complete engineering reports to keep tabs on. However, if you are like most developers, you will be very busy with other requirements and will appreciate the simplicity of a summary.

SITE ENGINEERING

Site engineering can become extremely expensive. The work is needed in most cases, but with proper management costs can usually be controlled. What types of issues are covered in the engineering of a site? Many factors are considered when engineering a site for development. Elements of the work cover everything from grading to site utilities. Roads, drainage, erosion, and other issues are looked at. Proper engineering, while expensive to pay for, can save you a great deal of money during the course of a development.

Grading and drainage are key issues in site engineering. Your experts have to design a plan that will ensure proper drainage and suitable control of storm water. This can involve anything from huge storm drains to simple gutters that empty into a subsurface drain. A development that is graded poorly and that has drainage problems is going to be a pain in the neck for the developer for many years to come.

Unless you are an engineer, you can't reasonably do your own engineering studies. You can look at topographical maps, review surveys, and eyeball your land, but you don't have the specific education or experience to make your own decisions for large projects. If you are developing only a couple of house lots, you might well be able to avoid the cost of site engineering. Common sense should tell you when a site is complex enough to warrant the services of an engineer.

The purpose behind site engineering is to create a development that is problem-free. While being a lofty goal, it is possible to design sites that will contain few flaws upon completion. An engineer will determine what the best means of providing access to a site is. Factors affecting access include:

- Safety
- Costs
- Topography

fastfacts

Drainage is a key element of a suitable building site. Don't cut corners when it comes to designing and creating proper drainage for projects that you are involved in. You may spend more money at the time of construction, but it will be money well spent to avoid headaches down the road.

> # **fast**facts
>
> *When you are assessing a building lot or a piece of land to develop, don't overlook the potential expenses that may be involved with the overall development.*

- Access permits
- General site design

Shaping the earth is part of land development. Filling in low spots, cutting out high spots, and grading for a uniform flow are all expected in most developing jobs. Developers will look for ways to limit the amount of fill that must be trucked into a site. By using natural land characteristics and building to suit them, developers can reduce the cost of earthwork. Maintaining the existing lay of the land is often beneficial, both in terms of limited disruption to the region and in cost savings.

Any risk of erosion or sedimentation must be considered during site engineering. Developments need utilities. How utilities will be brought to a site is often left up to the site engineer. The sizing of sewers and water mains is part of the engineering process. Storm sewers may be needed, as might fire hydrants. Routing the locations of utilities is another task to be dealt with. Engineers earn their money when it comes to large developments.

The handling of storm water can be a significant issue when developing land. There can be many rules affecting the treatment of storm water. In some cases, storm water will be piped to a storm sewer. Some cities allow storm water and sanitary sewage to share a common sewer, but many jurisdictions prohibit this practice. It's very common for a city to require two sewers, one for sanitary sewage and one for storm water. If there is no storm sewer available, what will happen with runoff water? Can you pipe it into a retention or detention pond? Perhaps, but there could be several regulations affecting how this might be done. Can a pipe be run on a gravity grade to a spot away from a building and then discharged onto open ground? In some places, yes, but other jurisdictions may frown upon this. Can underground dry wells (holes filled with gravel and covered with earth) be used to collect storm water? Probably, but you will have to check local code requirements.

When storm water is discharged into surface areas, there is concern for pollution of the local water table. Someone, you or your engineer, must determine the depth of the water table. County extension offices can usually provide data on the depth of the water table in various locations. Once the depth of the water table is known, you will need to know how much soil is required between the bottom of the discharge point and the water table. Local code requirements vary, so check with your local agencies to determine the depth requirements.

Soil types differ in their perk rates. You need to know the rates in order to make a decision about potential pollution. The bottom elevation of a catch basin, pond, or dry well is known as the invert. This is the point where measurement begins between the storm water and the water table. In other words, if you have a detention pond that is 8 feet deep and the water table is 12 feet deep, the distance between the storm water and the water table would be 4 feet. It's important that there be enough vertical distance between the water table and the surface water for the soil to cleanse the storm water before it enters the water table.

If your development work will include construction, your engineers will have to work with construction codes. Developers who are into conversion work, where buildings are bought, converted, and then sold or leased, also run into code issues. Meeting code requirements is a standard procedure for many of the experts involved with land development. Most local codes are well documented to eliminate confusion. Charts and tables are often provided in codebooks to aid planners.

GRADING DESIGNS

Grading designs for a development affect the control of storm water. Obviously, the lay of the land has a lot to do with the way that water runs. When you and your engineer are working on a grading plan, you should also be working on a storm-water plan. It makes sense to work on both concepts at the same time, since they are so tightly connected. The first step is the grading.

The less you have to alter existing land conditions, the better off you are. You will be saving time and money if you can work with the natural lay of the land. Developers who work with large parcels or who do extreme density usually have to cut and fill their projects. Smaller

fastfacts

Avoid buying land that is going to require a lot of expense in order to make it suitable for building. Most people will not be willing to pay a lot of additional money to obtain a normal building lot. So, if you have to invest substantial money to make a substandard parcel of land buildable, you may not see a good return on your investment.

developments, where building lots are larger and the density of housing is less, can sometimes avoid most of the cutting and filling.

You and your engineers have to look very closely at what your grading needs will be. If the work will be extensive, you can count on spending some serious money for work that most people will simply take for granted. If you invested tens of thousands of dollars in recreational amenities, such as a swimming pool and tennis courts, potential buyers would see value. Putting the same amount of money in dirt may be much more important, but buyers will not normally have much appreciation for your investment. It can be hard to recover your costs associated with earthwork.

When you have to cut and fill a site, you are exposing yourself to potential problems that would not exist if the ground alterations could be avoided. For example, when you cut away natural soil, you create a potential for erosion and sediment movement. Not only will this probably require an additional investment in containment control, but the process can slow your project down. Hauling in fill dirt is a very expensive proposition, and filled land often settles over time, which can create depressions that give the development drainage problems.

Sometimes the dirt cut from one section of a project can be used as fill for another piece of the project. While this method is not as nice as not having to cut at all, it is better than having to buy fill dirt and having it trucked into the site. Overall grading is a factor that should be weighed when assessing a piece of land for viability as a development project. Try to avoid land that will require extensive earthwork. If you can't avoid the need for cutting and filling, have your engineer design a plan that will minimize the work.

Before a grading plan can be completed, some benchmarks must be established. Engineers and surveyors can work together to establish desirable grades. Here are some examples of questions to consider:

- What will the finished grade level for housing be?
- How high above the finished grade level will the finished floor level of homes be?
- What will be required for parking areas and walkways?
- How will your roads be affected by various grades?

As your engineer answers these questions, you can begin to look for a spot elevation. This is a chosen elevation that will be used as a benchmark to meet your goals for all the other desired elevations. More than one spot elevation will be used to establish a network of measurements with which your engineer can work.

The engineering process for setting grades can be very complicated. However, the engineering reports are not difficult for most developers to understand. Let me give you a quick example. Assume that your surveyors have delivered an elevation survey to you and your engineer. The engineer has taken all spot elevations and plotted a house location on the survey. If you review the drawing, you can see almost exactly how the land around the home will affect drainage issues. Obviously, you want the land to slope away from the structure. But the slope must be within the required guidelines of local code restrictions for intended use. In this case, the intended use is a lawn. Your engineer will include notes on code requirements, recommended slope values, and actual slope values. You can see quickly how your plans fit the formula.

When you combine an elevation survey with spot elevations, you can determine many factors needed for construction. Surveyors can shoot elevations and mark them for all your development construction. For example, you can see on a grade stake just where the top of a road curb will be. The completed use of spot elevations may affect sewer depths, water service locations, finished floor grades, finished grade elevations, the depth of retention and detention ponds, required depths for sump pumps in basements, and so forth.

Doing a site grading study for a single house is fairly simple and easy to understand. But if you are doing the same work for a project that will contain 200 homes, the process becomes much more intricate. All the homes must be considered, as must all the open space, roads, parking areas, and other elements of the development. Designing a plan that will accommodate all the housing units and infrastructure takes experience, talent, and a lot of attention to

fastfacts

When you require engineering services, see professionals who specialize in the type of work that you need reports for. Many builders and developers go to a single firm where various types of engineers are employed. This is what I do, and I like the plan. Some builders seek out individual engineers from different companies. The choice is yours, but make sure that you have the right engineer for the job.

detail. This is when it feels very good to have competent engineers doing the math for you.

There are four types of calculations that may be used by a site engineer who is computing the needs of cutting and filling a site. If the site is small and will require little excavation, the prism method may be used. With this procedure, the engineer will multiply the area of excavations by the average height of the corners. In doing so, the engineer arrives at a figure to relate to the approximate volume of cutting or filling. Most developers should leave all final calculations to their experts.

The contour method is one where the area between existing and proposed contours are added together. Then the total is multiplied by the distance between the contours. The amount of cutting or filling is determined by the solution of the formula. While you may not be able to compute the needs for filling or cutting yourself, you should be able to understand the engineering report without much trouble.

Engineers may use the cross-sectional method to determine the needs for filling and cutting. They do this by drawing lines through a site plan to create sections. The sections are usually about 50 feet apart in scale. A planimeter, which is a tool that measures the areas of planes, is used to determine the planes. Each section is figured individually and then added together to arrive at the total requirements for cutting and filling.

Another method that might be used is the average-end method. A cross-section of an area reveals top and bottom areas as parallel planes. The two planes are added together and then divided by a formula of two multiplied by one. When done properly, the formula reveals the amount of cutting or filling that will be needed.

You don't have to worry about taking math classes at night to be a developer. It is not up to you to do your own engineering. But you should strive to understand as much of what you are seeing as you can. The fastest way to gain the knowledge is to question your engineer. If you ask a lot of questions, you may have to pay the engineers for their time in tutoring you. But simple questions should be answered quickly and probably without added cost. Learn what you can, where you can.

7

PRIVATE SEWAGE DISPOSAL SYSTEMS

There are site limitations for private sewage-disposal systems. Not all pieces of land are suitable for such systems. Land developers and builders must be able to spot land that is likely to give them trouble when it comes to a septic system. Experience does much to help in this area. After several years of buying and developing land, a person begins to know good soil when he or she sees it.

There are little signs that can give you hints about the quality of land. For example, bulges and occasional glimpses of rock on the land's surface could mean that bedrock is close to the surface. This will certainly interfere with a private sewage-disposal system.

Rock is not the only risk when it comes to septic systems. Some ground just won't perk. If this is the case, the land may be deemed unbuildable. Buying land for development and then discovering that there are no acceptable septic sites can really ruin your day. Experienced land buyers have clauses in their purchase agreements to protect them from this type of risk. A typical land agreement will contain a contingency clause that gives the buyer a chance to have the soil tested before an absolute commitment is made to purchase it. If the tests prove favorable, the deal goes through. When soil studies turn up problems, the contract might be voided or some compensation might be made in the sales price.

The size of a building lot can affect its ability to be approved for a septic system. Many houses that require a septic system for sewage disposal also require a water well for drinking water. For

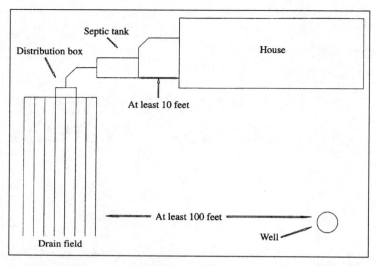

FIGURE 7.1 Septic layout showing recommended minimum distance from a water well.

obvious reasons, septic systems must not be installed too close to a water well. The minimum distance between the two is normally 100 feet or more. I have seen some exceptions to this rule but not many.

I've run into a number of building lots where municipal water hook-ups were available but where a private waste disposal system was needed. With my contingency clauses and inspections, I've never been put in a bind by this type of problem. But a buyer or

fastfacts

Never assume that you will not have trouble finding a site for a private sewage system just because a parcel of land is large or looks good. Until you have a soils study done, you can't be sure that the land will support a private septic system. However, with modern technology, almost all reasonable land can support a private sewage system if you are willing to invest enough money in it.

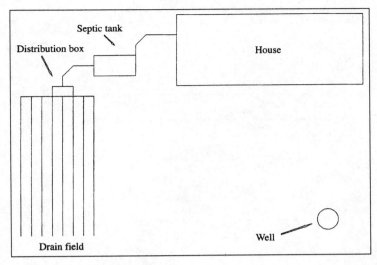

FIGURE 7.2 Typical septic layout.

builder who doesn't research what options are available could get in big trouble very quickly.

Some people assume that since building lots on either side of a particular lot are able to use septic systems, the lot in question should be suitable for such a system. This is not always the case. I've seen land where out of five acres there might be only one or two sites suitable for a standard septic system. This situation is not uncommon, so watch out for it. Make sure that you have an approved septic location established before you make any firm commitments to buy or build.

GRADE

The grade of a building lot can have a great deal to do with the type of septic system that must be installed. If the grade will not allow for a gravity system, the price that you pay will go up considerably. Pump stations can be used, but they are not inexpensive. If you're building houses on spec, you might have trouble selling one that relies on a pump station. People don't like the idea of having to replace pumps at some time in the future. And many people are

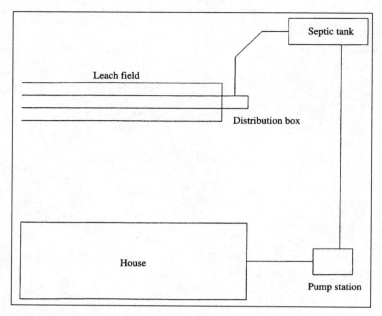

FIGURE 7.3 Septic system with pump system.

afraid that the pumps will fail, leaving them without sanitary conditions until it can be replaced.

The naked eye is a natural wonder, but it cannot always detect the slope of a piece of land accurately. Looks can be deceiving. Unless the land you are looking at leaves no room for doubt about its elevation, check the land with a transit. You can't afford to bid jobs with regular septic systems and then wind up having to install pump systems.

SEPTIC-SYSTEM DESIGN

There is a safe way to work with land where a septic system is needed. You can go by reports that are provided to you by experts. Even this is not foolproof, but it's as good as it gets. If you have a soils engineer design a septic system for your job, you can be pretty sure that the system will work out close to the way it has been drawn. Since every jurisdiction I know of requires a septic design before a building permit will be issued, it makes sense to go ahead and get the design early.

If you are buying land, make sure that your purchase agreement provides a contingency that will allow you to have the land approved for a septic system before you are committed to going through with the sale. It is wise to specify in your contingency what type or types of septic designs you will accept. Some types of systems, such as chamber systems, cost much more than others. If your contract merely states that the land must be suitable for a septic system, you will have to complete the sale regardless of what the cost of a system is, so long as one can be installed.

When you are bidding jobs, make clear in your quote what your limits are in regard to site conditions. Specify the type of septic system your quote is based on, and provide language that will protect you if some other type of system is required. Don't take any chances when dealing with septic systems, because the money you lose can be significant.

STANDARD SYSTEMS

A standard septic system will be built with a gravity flow. Its drain field—or leach field, as it is often called—will be made up of perforated pipe and crushed stone. This is the least expensive type of septic system to install. Unless the soil will not perk well, you will normally be able to use a standard system.

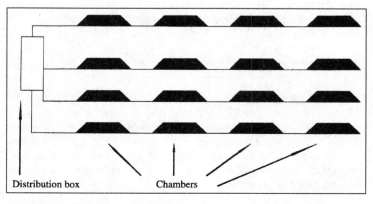

FIGURE 7.4 Chamber-type septic system.

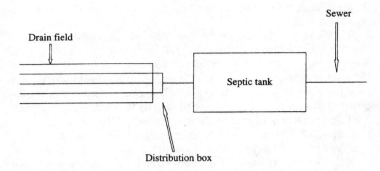

FIGURE 7.5 Components of a septic system.

CHAMBER SYSTEMS

Chamber systems are much more expensive than pipe-and-gravel systems. When soil doesn't perk well, chamber systems are used. If the soil will perk but not well, a chamber system may be your only choice. How do these systems work? Basically, the chambers hold effluent from a septic tank until the ground can absorb it. Unlike a perforated pipe, which would release effluent quickly, the chamber controls the flow of effluent at a rate acceptable to the soil conditions. Where a pipe-and-gravel system might flood an area with effluent, a chamber system can distribute the liquid more slowly and under controlled circumstances. The cost of a chamber system can easily be twice that of a pipe-and-gravel system.

PUMP STATIONS

Pump stations are another big expense in some septic systems. The cost of the pump station, the pump and its control, and the additional labor required to install such a system can add thousands of dollars to the cost of a standard septic system. If you were unlucky enough to get stuck with a pumped chamber system when you had planned on a gravity gravel system, you could lose much of your building profit all at once.

TREES

Trees are another factor you must consider when doing a site inspection for a septic system. Tree roots and drain fields don't mix very well. If there are trees in the area of the septic system, they must be removed. Even trees that are not directly in the septic site should be removed from the edges of the area. How far should open space exist between a septic system and trees? It depends on the types of trees that are growing in the area. Some trees have roots that reach out much further than others. You can get a good idea of how far a tree's roots extend by looking at the branches on the tree. The spread of the branches will often be similar to the spread of the roots. If there is any doubt, ask the professional who draws your septic design to advise you on which trees can be left standing.

BURYING A SEPTIC TANK

Burying a septic tank requires a fairly deep hole. Even if you are using a low-profile tank, the depth requirement will be several feet. If you are working in an area where bedrock is present, you must be cautious. You could run into a situation where rock will prevent you

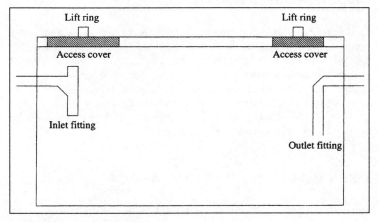

FIGURE 7.6 Common septic tank.

TABLE 7.1   Common septic tank capacities.

Single family dwellings, number of bedrooms	Multiple dwelling units or apartments, one bedroom each	Other uses; Maximum fixture units served	Minimum septic tank capacity in gallons
1–3		20	1000
4	2	25	1200
5 or 6	3	33	1500
7 or 8	4	45	2000
	5	55	2250
	6	60	2500
	7	70	2750
	8	80	3000
	9	90	3250
	10	100	3500

from burying a septic tank. If this is the case, you can usually haul in dirt and design a system that will work.

UNDERGROUND WATER

Underground water can present problems for the installation of a septic system. This type of problem should be detected when test pits are dug for perk tests. However, it is possible that the path of the water would evade detection until full-scale excavation was started. For this reason, you should have some type of language in your contracts with customers to indemnify you against underground obstacles such as water. If water is a problem, it should be able to be overcome with a mound system.

DRIVEWAYS AND PARKING AREAS

When you assess a lot for a septic system, consider the placement of driveways and parking areas. Even though a septic system will be below ground, it is not wise to drive vehicles over it. The weight and movement can damage the drain field to a point where replacement is required. You certainly don't need this type of warranty work, so make sure that all vehicular traffic will avoid the septic system.

EROSION

Erosion can be a problem with some building lots and land. If you install a septic field on the side of a hill, you must make sure that the soil covering the field will remain in place. This can be done by planting grass or some other ground cover. When you check out a piece of land, you need to take the erosion factor into consideration. The cost of preventing a washout over the septic system could add a significant amount of expense to your job.

SETBACKS

Setbacks are another factor that you should check on before committing to a septic design. Many localities require all improvements made on a piece of land to be kept a minimum distance from the property lines. A typical setback for a side property line is 15 feet, but this is not always the case. Where I live, there are no general setbacks for building, but a septic system must be at least 10 feet from an adjoining property line. I've seen setback requirements that were more than 15 feet. This can become a very big factor in the installation of a septic system. As always, check your local requirements prior to putting people to work.

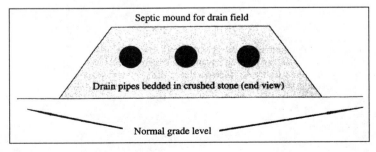

FIGURE 7.7 Mound-type septic system

PRIVATE WATER SUPPLIES

Site evaluations for water wells are often taken for granted by builders. I have known many builders who submitted bids for work without ever seeing the building lot. There is a lot of risk in doing this. A builder cannot bid a job competitively without knowing what type of well will be used. If a bid goes in for a bored well and it turns out that a drilled well is needed, the bidder will lose money. When compensation for not conducting a site inspection is made by specifying the most expensive type of well, the job can be lost by being bid too high. Failure to do a site inspection can be a very big mistake.

You can't always see what's likely to be under the ground by looking at the surface. Knowledgeable builders want to know what they will get into when drilling wells and digging footings. Many experienced builders require customers or landowners to provide them with soil studies before giving a firm bid. When such studies have not been done, some builders do their own. I'm one of these builders.

It is not uncommon to see me out digging holes on potential building sites. A post-hole digger can reveal a lot about what conditions exist below the topsoil. Augers and probe rods can also provide some insight into what is likely to be encountered. A probe rod will tell you if a lot of rock is present. But, to see the soils, you need a hole. An auger or post-hole digger is the best way to get these samples. Augers are often easier to use; a power auger is ideal.

When you create test holes, you have a lot more information to base your bid decisions on. There are only a few ways to bid a job where a well will be installed. You can guess what will be needed, but this is very risky. Digging test holes will give you a very good idea of what types of wells might be suitable. Interviewing well owners on surrounding property can provide a lot of data that can help you with your decision. And hiring soil-testing companies is a great, but expensive, way to find out what you are getting into. Having a few well installers walk the land with you so that they can provide you with solid bids is another good way to protect yourself.

LOCATION

The location of a well is important. Choosing a location is not always easy. Many factors can influence the location of a well. The most obvious might be the location of a house. It is not common to place a well beneath a home, so most people will choose a location outside of the foundation area of a home. Septic fields are another prime concern. Wells are required to be kept a certain distance from septic systems. The distance can vary from jurisdiction to jurisdiction and because of topography. Access to a location with a well-drilling rig is also a big factor. These large trucks aren't as maneuverable as a pick-up truck. Picking a place for a well must be done with access in mind.

Finding suitable sites for a well can be made easier with some background information and site inspections. Maps can give you a lot of guidance on where water might be found. Some regional authorities maintain records on wells already in existence. Reviewing this historical data can definitely help you pinpoint your well type and location. Unfortunately, there is never any guarantee that water will be where you think it is. A neighboring landowner might have a well that is 75 feet deep while your well turns out to be 150 feet deep. It is, however, likely that wells drilled in close proximity will average about the same depth. I've seen houses in subdivisions where one house has great well water and the next-door neighbor's water suffers from unpleasant sulfur content. 50 feet can make quite a difference in the depth, quantity, and quality of a well.

Plants

Plants can be very good indicators of what water is available beneath the ground's surface. Trees and plants require water. The

fastfacts

If you have any doubt about how deep water will be found, plan and budget for a drilled well. Most well drillers will give you a flat-rate price for a drilled well that guarantees water. Sometimes this might be more expensive than what you might have been able to contract, but at least you will have water at a known price.

fact that plants and trees need water might not seem to provide much insight into underground water. Take cattails as an example. If you see cattails growing nearby, you can count on water being close by. It's suggested that the depth of water in the earth can be predicted, to some extent, by the types of trees and plants in the area.

Cane and reeds are believed to indicate that water is within 10 feet of the ground's surface. Arrow weed means that water is within 20 feet of the surface. There are many other guidelines for various types of plants and trees in regard to finding water. From a well-drilling point-of-view, I'm not sure how accurate these predictions are. I know that cattails and ferns indicate water is close by, but I can't say that it will be potable water or how deep it will be found in the ground. I suspect that there are some very good ways to predict water with plants and trees.

Since I am not an expert in plants, trees, or finding water, I won't attempt to pump you up with ideas of how to find water at a certain depth just because some particular plant grows in the area. It is my belief that with enough knowledge and research, a person can probably predict water depths with good accuracy, in many cases.

DRILLED WELLS

Drilled wells are the most dependable individual water source I know of. These wells extend deep into the earth. They reach water sources that other types of wells can't come close to tapping. Since drilled wells take advantage of water that is found deep in the ground, it is very unusual for drilled wells to run dry. This accounts for their dependability. However, dependability can be expensive.

Of all the common well types, drilled wells are the most expensive to install. The difference in price between a dug well and a drilled well can be thousands of dollars. But the money is usually well spent. Dug wells can dry up during hot summer months. Contamination of well water is also more likely in a dug well. When all the factors are weighed, drilled wells are worth their price.

Depth

The depth of a drilled well can vary a great deal. In my experience, drilled wells are usually at least 100 feet deep. Some drilled wells extend 500 feet or more into the earth. My personal well is a little over 400 feet deep. Based on my experience as both a builder and a plumber, I've found most drilled wells to range between 125 to 250 feet deep. When you think about it, that's pretty deep.

It's hard for some people to envision drilling several hundred feet into the earth. I've had many homebuyers ask me how I was going to give them such a deep well. Many people have asked me what will happen if the well driller hits rock. Getting through bedrock is not a problem for the right well-drilling rig. While we are on the subject of drilling rigs, let's talk about the different ways to drill a well.

Well-Drilling Equipment

Well-drilling equipment is available in various forms. While one type of rig may be the most common, all types of well rigs have advantages and disadvantages. As a builder, it can be helpful to know what your options are for drilling wells on different types of sites. There are two basic types of drilling equipment.

Rotary drilling equipment is very common in my area. This type of rig uses a bit to auger its way into the earth. The bit is attached to a drill pipe. Extra lengths of pipe can be added as the bit cuts deeper into the ground. The well hole is constantly cleaning out air, water, or mud under pressure.

Percussion cable tool rigs make up the second type of drilling rig. These drilling machines use a bit that is attached to a wire cable. The cable is raised and dropped repeatedly to create a hole. A bailer is used to remove debris from the hole. Between rotary equipment

fastfacts

Make sure any well you provide is disinfected prior to use. Some drillers do this work, but it is usually up to the plumber or the builder. Never allow a well to be used prior to disinfection and subsequent water test to determine that the water is safe for human consumption.

and percussion cable equipment, there are a number of variations in the specific types of drilling rigs.

We could go into a lengthy discussion of all the various types of well rigs available. But since you are a builder and probably have no desire to become a well driller, there seems to be little point in delving into all the details of drilling a well. What you should know, however, is that there are several types of drilling rigs in existence. Your regional location may affect the types of rigs being used. A few phone calls to professional drillers will make you aware of what your well options are. For my money, I've usually contracted the services of rotary drillers. There are times when other types of rigs could be a better choice. My best advice to you is to check with a number of drilling companies in your area and see what they recommend.

THE BASICS

Let's go over the basics of what is involved, from a builder's point of view, when it comes to drilling a well. Your first step must be deciding on a well location. The site of a proposed house will, of course, have some bearing on where you want the well to be drilled. Local code requirements will address issues pertaining to water wells. For example, the well will have to be kept at some minimum distance from a septic field, assuming that one is to be used.

Who should decide on where a well will go? Once local code requirements are observed, the decision for a well location can be made by a builder, a home buyer, a well driller, or just about anyone else. If you're building spec houses, the decision will be up to you and your driller. Buyers of custom homes may want to take an active role in choosing a suitable well site. As an experienced builder, I recommend that you consult with your customers on where they would like their well. Some people are adamant about where they do and do not want a well placed.

Access

Access is one of the biggest concerns a builder has in the well-drilling process. It is a builder's responsibility to make access available to a well driller. Drilling rigs require a lot of room to maneuver. A narrow, private drive with overhanging trees may not be suitable for a well rig.

Having enough width and height to get a well truck into a location is not the only consideration. Well rigs are heavy—very heavy, in fact. The ground that these rigs drive over must be solid. New construction usually requires building roads or driveways. If you can arrange to have a well installed along the roadway, your problems with access are reduced.

It is not always desirable to install a well alongside a driveway. This may not cause any additional concern, but it could. If the ground where you are working is dry and solid, a well rig can drive right over it. But, if the ground is wet and muddy or too sandy, a big truck won't be able to cross it. You must consider this possibility when planning a well installation.

Don't build obstacles for yourself. Just as the old joke goes about painting yourself into a corner, you can build yourself right out of room. Some builders put wells in before they build houses. Others wait until the last minute to install a well. Why do they wait? They do it to avoid spending the money for a well before it's needed. This reduces the interest they pay on construction loans and keeps their operating capital as high as possible. When the well is installed last, it can often be paid for out of the closing proceeds from the sale of a house.

fastfacts

It is a good idea to install a well prior to home construction. I used to save the well work until the end of projects for many years. It was a way of keeping the outstanding balance on the construction loan down. Then, somewhere along the way, I realized that I might build a house and not be able to find water. After that I decided to drill the wells early in the construction process.

I have frequently waited until the end of a job to install wells. There is some risk to this method. It's possible, I suppose, that a house could be built on land where no water could be found with a well. This would truly be a mess. A more likely risk is that the house construction will block the path of a well rig. In my later, more experienced years, I have installed the wells prior to home construction.

When you are confirming the location for a well, make sure that you will be able to get drilling equipment to it when you need to. Going to the buyers of a custom home and informing them that their beautiful shade trees will have to be cut down to get a well rig into the site is not a job I would enjoy having. It is safer to install wells before foundations are put in.

Working on Your Site

Once you have a well driller working on your site, there's not much for you to do but wait. The drilling process can sometimes be completed in a single day. There are times, however, when the rig will be working for longer periods of time. Depending on the type of drilling rig being used, a pile of debris will be left behind. This doesn't usually amount to much, and your site contractor can take care of the pile when preparing for finish grading.

The well driller will usually drill a hole that is suitable for a 6-inch steel casing. The casing will be installed to whatever depth is necessary. Once bedrock is penetrated, the rock becomes the well casing. How much casing is needed will affect the price of your well. Obviously, the less casing that is needed, the lower the price should be.

Your well driller will grout the well casing as needed to prevent ground water from running through the casing and dropping down the well. This is normally a code requirement. It limits the risk of contamination in the well from surface water. A metal cap is then installed on the top edge of the casing, and the well driller's work is done.

Pump Installers

Many well drillers offer their services as pump installers. You might prefer to have the driller install the pump, or maybe you would rather have your plumber do the job. A license may be needed in your area for pump installations. Any master plumber can install a pump, but you should check to see if drillers are required to have special installation licenses. If they are, make sure that any driller you allow to install a pump is properly licensed.

fastfacts

Many jurisdictions require the wiring for a well pump to be installed by a licensed electrician. Plumbers and well drillers did this type of work for years, but some jurisdictions are now requiring the work to be done by licensed electricians. Check your local code requirements if you have any doubts on this issue.

If you want your driller to install a pump set-up as soon as the drilling is done, you will need to make arrangements for a trench to run the water service pipe in. The cost of this trench is usually not included in the prices quoted by pump installers. Watch out for this one, because it is an easy way to lose a few hundred dollars if you don't plan on the cost of the trench.

After your well is installed, you should take some steps to protect it during construction. Heavy equipment could run into the well casing and damage it. I suggest that you surround the well area with some highly visable barrier. Colored warning tape works well, and it can be supported with nothing more than some tree branches stuck in the ground. A lot of builders don't take this safety precaution, but they should. A bulldozer can really do a number on a well, and a well casing can be difficult for an operator to see at times.

How far will the pump installer take the job? Will the water service pipe be run just inside the home's foundation and left for future hookup? Who will run the electrical wires and make the electrical connections? Does the installation price include a pressure tank and all the accessories needed to trim it out? Who will install the pressure tank? You need answers to these questions before you award a job to a subcontractor. Pump systems involve a lot of steps and materials. It's easy for a contractor to come in with a low bid by shaving off some of the work responsibilities in fine print. Be careful, or you may wind up paying a lot more than you planned to for a well system.

QUANTITY AND QUALITY

When it comes to the quantity and quality of water produced by a well, few (if any) well drillers will make commitments. Every well driller I've ever talked to has refused to guarantee the quantity or quality of water. The only guarantee that I've been able to solicit has been one of hitting water. This is an issue that builders have to be aware of.

Customers may ask you to specify what the flow rate of their new well will be. It is beyond the ability of a builder to do this. An average, acceptable well could have a 3-gallon-per-minute recovery rate. Another well might replenish itself at a rate of 5 gallons per minute. There are wells with much faster recovery rates, and there are others with slower rates. Anything less than 3 gallons per minute is less than desirable, but it can be made to suffice. The key to making a well with a slow recovery rate work is to have it deep enough to maintain a sufficient supply of water at static level.

How can you deal with the recovery rate? You really can't. Sometimes by going deeper, a well driller can hit a better aquifer that will produce a higher rate of recovery, but there is no guarantee. Don't give a guarantee to your customers. Show your customers the disclaimers on the quotes from well drillers and use that evidence to back up your point that there is no guarantee of recovery rate in the well business. Now I could be wrong. You might find some driller who will guarantee a rate, but I never have.

Even though you can't know what the recovery rate will be when you start to drill a well, you can determine what it is after water has been hit. Your well driller should be willing to test for and establish the recovery rate for you. Every driller I've ever used has performed this service. You need to know what the recovery rate is in order to size a pump properly. There is an old-fashioned way of determining a recovery rate, but I'd ask my driller to do it for me.

It is also important that you know the depth of the well, and this is something that your driller can certainly tell you. The depth of the well will also be a factor when selecting and installing a pump. Don't let your driller leave the job until you know what the depth and the recovery rate are.

Quality

The quality of water is difficult to determine when a new well is first drilled. It can take days, or even weeks, for the water in a new well to assume its posture. In other words, the water you test today may offer very different results when tested two weeks from now. Before a true test of water quality can be conducted, it is often necessary to disinfect a new well. Many local codes require disinfection before testing for quality. Even if they don't, you should insist on it.

Wells are usually treated with chlorine bleach to disinfect them. Local requirements on disinfection vary, so I won't attempt to tell you exactly what to do. In general, a prescribed amount of bleach is poured into a well. It is allowed to sit for some specific amount of time, as regulated by local authorities. Then the well pump is run to deplete the water supply in the well. As a rule of thumb, the pump is run until there is no trace of chlorine odor in the tap water. When the well replenishes itself, the new water in the well should be ready for testing. But, again, check with your local authorities for the correct procedure to use in your area. By the way, builders usually are responsible for conducting the disinfection process.

Once the well is ready to test, a water sample is taken from some faucet in the house. Test bottles and collection instructions are available from independent laboratories. Follow the instructions provided by the testing facility. As a rule of thumb, you should remove any aerator that may be installed on the spout of a faucet before taking your water collection. It is often recommended that a flame be held to the spout to kill bacteria that may be clinging to the faucet and washed into the collection bottle. Don't attempt this step when testing from a plastic faucet! Many plumbers take water tests from outside hose bibbs, and this is fine. A torch can be used to sterilize the end of the hose bibb, which provides almost direct access to water.

When collecting water for a test, water should be run through the faucet for several minutes before catching any for a test. It is best to drain the contents of a pressure tank and have it refill with fresh well water for the test. You can run reserve water out quickly by opening the cold-water faucet for a bathtub.

fastfacts

When collecting water for a water test, let the water run for a few minutes before catching your sample. Let the water that has been lying in the pipes evacuate prior to the capture. It is recommended to sterilize the end of the outlet where the water is to be caught. You can do this with a torch, a few matches, or a cigarette lighter. If you use a torch, don't melt the fixture outlet.

Once a water sample is collected, it is taken or mailed to a lab. Time is of the essence when testing for bacteria, so don't let a water bottle ride around in your truck for a few days before you get around to mailing it. Again, follow the instructions provided by the lab for delivering the water.

There are different types of tests that you can request the lab to perform. A mandatory test will reveal if the water is safe to drink. If you want to know more, you have to ask for additional testing. For example, you might want to have a test done to see if radon is present. Many wells have mineral contents in sufficient quantity to affect the water quality. Acid levels in the well may be too high. The water could be considered hard, which can make washing with soaps difficult and stain plumbing fixtures. There are a number of potential tests to run.

Drinking water, or potable water as it can be called, is your primary goal when drilling a well. It is rare that a drilled well will not test well enough to meet minimum requirements for safe drinking. In fact, I've never known of a drilled well that wasn't suitable for drinking. Nevertheless, an official statement of acceptability is generally required by code officials and lenders who loan money on houses.

If you get into a discussion on water quality, you may have to pinpoint exactly what it is that you are talking about. Are you only talking about the water being safe to drink? Or does your discussion include mineral contents and such? Most professionals look at water quality on an overall basis, which includes mineral contents.

TRENCHING

Trenching will be needed for the installation of a water service and for the electrical wires running out to a submersible pump. It is possible to pump water from a deep well with a two-pipe jet pump, but submersible pumps are, in my opinion, far superior. When a submersible pump is used, it hangs in the well water. Electrical wires must be run to the well casing and down into the well. The wires and the water service pipe can share the same trench.

As with any digging, you must make sure that there are no underground utilities in the path of your excavation. Most communities offer some type of underground utility identification service. In many places, one phone call will be all that is needed to get all underground utility locations on your work site marked. It may, however, be necessary in some parts of the country to call individual utility companies. I expect that you are familiar with this process; most builders are.

Once you have a clear path to dig, a trench must be dug to a depth that is below the local frost line. The water service pipe will have water in it at all times, so it must be buried deep enough to avoid freezing. How deep is deep enough? It varies from place to place. In Maine, I have to get down to a depth of 4 feet to be safe from freezing. In Virginia, the frost line was set at 18 inches. Your local code office can tell you what the prescribed depth is in your area.

After a trench is opened up, you can arrange for your pump installer. Most installers will want to do all of their work in one trip. This means that you must have enough of the house built to allow an installer to bring the water pipe through the foundation and to set the pressure tank. As a helpful hint, you should install a sleeve in your foundation as it is being poured so that the pump installer will not have to cut a hole through the foundation.

Check with your local plumbing inspector to determine what size sleeve will be needed. Most plumbing codes require a sleeve in a foundation wall to be at least two pipe sizes larger than the pipe being installed. For a typical 1-inch well pipe, this would mean that the sleeve would have to be at least 2 inches in diameter. But, again, check with your local code office, because plumbing codes do vary from place to place.

SITE WORK
AND SITE UTILITIES

Only the weather can be counted on to provide as many unknowns as those encountered when significant amounts of site work are undertaken at the construction site. Unforeseen subsurface conditions serve as the basis for many construction-related disputes and claims.

During the initial stages of the design of a project requiring new foundations or any extensive subsurface excavation, a civil engineer will make arrangements to take test borings in order to determine the nature, composition and load-bearing capacities of the soils in questions. Figure 9.1, is a representative log of a test boring indicating the depth of the boring and the soil conditions encountered until reaching its final depth of 74.3 feet. The geotechnical report (Figure 9.2), prepared after an analysis of soil conditions has been made, will generally contain a narrative of the exploration process accompanied by test boring reports. These reports provide definitive information about the soils in the area where they are taken; but since soil strata can change dramatically in a relatively short distance, the test borings can act only as "guidelines" and not as definitive documentation of the conditions to be encountered over the entire area.

UNDERSTANDING THE NATURE OF SOIL

A better understanding of the nature and types of soil conditions is essential to learn how to deal with the many problems that may

LOG OF BORING No. B-1

CLIENT: ATLAS SYSTEMS, INC.		DATE: 6-22-99	#02995604	RIG: CME 75
SITE: 1026B South Powell Road Independence, Missouri		PROJECT: ATLAS LOAD TEST		

GRAPHIC LOG	DESCRIPTION	DEPTH ft.	USCS SYMBOL	NUMBER	TYPE	RECOVERY in.	SPT – N BLOWS/ft	WATER CONTENT %	DRY UNIT WT. pcf	UNCONFINED STRENGTH Qu psf	ATTERBERG LIMITS LL, PL, PI
	6" GRAVEL LEAN CLAY, silty trace organics, gray brown, trace dark brown and red brown, medium (Possible Fill)				PA						
			CL	1	SS	14	7	34.1		2000*	45,21,34
		5			HS						
	LEAN CLAY, calcareous, trace sand and limestone gravel dark brown, brown, very stiff (Possible Fill)										
		10	CL	2	SS	6	5	18.6		7000*	45,23,22
					HS						
		15	CL	3	SS	24	9	24.1		5500*	
					HS						
	LEAN CLAY, trace silt, gray brown, trace dark gray, red brown and dark brown, stiff to very stiff										
		20	CL	4	SS	24	10	22.3		3500*	44,20,24
					HS						
		25	CL	5	SS	24	5	27.6		2500*	
					HS						
	LEAN CLAY, silty, gray brown, trace dark brown, stiff to very stiff										
		30	CL	6	SS	24	19	26.5		5000*	42,18,24
					HS						
	Trace limonites at 34.0'	35	CL-CH	7	SS	24	14	23.5		5000*	
					HS						
	LEAN TO FAT CLAY, gray brown, trace dark brown, very stiff										

FIGURE 9.1 Typical test boring.

GRAPHIC LOG	DESCRIPTION	DEPTH ft.	USCS SYMBOL	NUMBER	TYPE	RECOVERY in.	SPT – N BLOWS/ft	WATER CONTENT %	DRY UNIT WT. pcf	UNCONFINED STRENGTH qu psf	ATTERBERG LIMITS LL, PL, PI
	LEAN TO FAT CLAY, gray brown, trace dark brown, very stiff	40	CL-CH	8	SS	24	13	24.3		5000*	48,20,26
					HS						
	LEAN CLAY, silty, gray brown, brown, trace dark brown, red brown and gray, medium to stiff	45	CL	9	SS	24	11	24.5		3000*	
					HS						
	Trace gravel at 49.0'										
		50	CL	10	SS	24	10	26.3		3000*	46,21,25
					HS						
		55	CL	11	SS	24	13	24.7		3500*	
					WB						
	Trace gravel at 59.0'										
		60	CL	12	SS	24	9	25.7		1500*	
					WB						
	***SHALE, highly weathered, trace silty clay and gravel olive brown, gray trace brown	65		13	SS	16	55	21.1		9000*	
					WB						
	***SHALE, highly weathered, trace clay and black coal, very dark gray, gray	70		14	SS	6	44/3"	37.6			
					WB						
	***SHALE, highly weathered, calcareous, gray 74.3 BOTTOM OF BORING			15	SS		50/1"	13.1		3000*	
WATER LEVEL OBSERVATIONS, ft. NONE – WD NONE - AB											

* Calibrated Hand Penetrometer
** CME 140H SPT automatic hammer
*** Classification estimated from disturbed samples. Core samples and petrographic analysis may reveal other rock types.
The stratification lines represent the approximate boundary lines between soil and rock types: in-situ, the transition may be gradual.

FIGURE 9.1 Typical test boring *(continued)*.

Geotechnical Report

The geotechnical report provides a summary of the findings of the three phases detailed above and also contains recommendations on options for a foundation together with the recommended soil related design values. Included in this report are the results of the laboratory testing of the soil samples and borings logs providing a visual summary of the vertical profile of foundation soils at the project site. *Figure 4* gives the boring log generated from the field exploration program as shown in *Figures 1* through *3*. A review of this boring log indicates the following:

- The total depth of the boring was 74.3 ft. Except for the upper one-half foot, the soil layers were all lean or lean to fat clay with some variations in color and stiffness down to the depth of 64 ft. At 64 ft., a shale stratum was encountered which was in a highly weathered condition. Shale is a rock but in a weathered condition such as noted on the boring log, it is probable that the helical plates of a pier could be set 1 ft. to 3 ft. into the shale stratum.

- Standard Penetration Tests (SPT) were conducted at each 5-ft. interval of depth down to the bottom of the boring. From the SPT, N column on the boring log, it is noted that the stiffness (or strength) of the lean clay is fairly consistent from depth 30 ft. to 64 ft. (N ranged from 9 to 14). The upper part of this stratum (around 35 ft. to 40 ft. is where the helical plates would be seated).

- Moisture contents were taken on the recovered split spoon samples from the SPT. Again, below about 25 ft., the moisture content of the soil was fairly consistent (ranging between 23-1/2 to 26-1/2 percent). This low variation in moisture content is consistent with the consistent range of N values.

- Liquid Limit and Plastic Limit tests were also conducted on the recovered split spoon samples. The average LL = 45; the average PL = 20, resulting in a PI = 25. These results indicate that the in-situ moisture content of the lean clay ($\cong$ 25%) from 30 ft. to 60 ft. is just above the Plastic Limit (20%). As the in-situ moisture content approaches the Plastic Limit, the clay soil will become stiffer (higher cohesion).

- The boring log also shows a column for unconfined compression strength for the clay type soil. The values indicated in this column were determined in the field using a hand held penetrometer device inserted into the split spoon sampler at the end of the exposed soil sample. This is not recognized as an accurate type of strength test for clay type soil but does provide a general order of magnitude of strength and also allows a comparison of strengths between various depths in the boring log. At a depth of 35 ft., the unconfined compression strength, q_U = 5000 psf. The cohesion of the soil, c, is taken to be 1/2 of qu or c = 2500 psf. This is generally a higher value than would be determined through correlation of the cohesion through the N value. From standard correlation charts, the cohesion, c, would likely be between 1500 and 1800 psf.

- In the bottom left hand corner of the boring log of *Figure 4*, it is indicated that no ground water table (GWT) was identified at the site.

- The fourth column in *Figure 4* shows the soil classification symbols to help describe the soils. *Table 2*, Page A10 gives an abbreviated listing of the Unified Soil Classification system.

FIGURE 9.2   The geotechnical report.

occur during site work and site utility operations. When reading a soils report or specification issued by a geotechnical consultant or a civil engineer, terms and designations require some interpretation for those who do not deal with soil composition frequently.

Soils are identified and classified according to particle size and fall into four major groups:

- Clay: Particle sizes 0.00024 inch (.006 mm) or smaller
- Silt: Particle sizes ranging from 0.00024 inch (.006 mm) to 0.003 inch (.076 mm)
- Sand: Particle sizes ranging from 0.003 inch (.076 mm) to 0.08 inch (2.03mm)
- Gravel: Particle sizes ranging from 0.08 inch (2.03mm) to 3 inches (76.2mm)

There are three basic soil classification systems:

- Tyler system: A method of determining particle size by using openings per lineal inch of wire screen; for example, a No. 20 mesh has 20 openings per lineal inch
- Unified Soil Classification System: This is the most widely used and accepted system; it uses letters to designate soil types within three major groupings: coarse grained, fine grained, and highly organic soils
- OSHA Soil Classification System: Although this system is not used by the engineering profession it is used to determine whether or not the soils present in trench excavation will require various slope configurations to remain stable or are of such a nature that a trench box shoring will be required to provide safe working conditions in the trench and avoid cave-ins.

fastfacts

Injuries and fatalities from excavation cave-ins are the fifth most common accident in the construction industry according to OSHA records.

The Unified Soil Classification (USC) System

The USC system of soil classification is divided into three major divisions, each with several subcategories and each identified by a symbol:

1. Coarse-grained soils: This category includes gravel, sand, and mixtures of both types.The USC letter designation for these soils is "G" for gravel and "S" for sand. When mixtures are classified, they will be designated either "GS" or "SG" according to which category, sand or gravel, is the primary ingredient. Sand and gravel are further categorized into four subgroups:

   • Well graded: Designated by the letter "W"

   • Poorly graded: Designated by the letter "P"

   • Dirty with plastic fines: Also designated by the letter "P"

   • Dirty with nonplastic silty fines: Designated by the letter "W" when this material will pass through a No. 200 sieve

2. Fine-grained soils: These types of soils are further categorized into soils (M), inorganic clays (C), and organic silts or clay (O), and each of these groups is further divided into soils having liquid limits (water-holding capacity) lower than 50 percent (L) and those with liquid limits higher than 50 percent (H). For example, a soil with inorganic silt with a liquid limit lower than 50 percent would be designated (ML).

3. Highly organic soils: This group is identified by the letters Pt for peat which is a characteristic of this group.

The OSHA Method of Classifying Soils

OSHA is primarily concerned with the composition of soil and how it relates to the stability of excavated trench walls. The OSHA soil classification is based largely upon the degree of cohesiveness of soil for safety purposes. They classify soils in four categories: stable rock, A, B, and C soils in decreasing order of stability; Type A is the most stable and Type C the least stable, requiring the maximum slope when trenching:

• Rock: Natural solid material that can be excavated with vertical sides that will remain intact while exposed

• Type A: Clay, silty clay, sandy clay, clay loam

• Type B: Silt, silt loam, sandy loam

- Type C: Granular soils such as gravel, sand, loamy sand, and soil from which water is freely seeping

OSHA dictates the maximum allowable slopes in terms of the allowable angle from a vertical plane:

Soil Type	Maximum Allowable Slope for Excavations less than 20 Feet (6 meters) deep
Stable Rock	Vertical (90 degrees from the horizontal)
Type A	Maximum allowable slope for excavation less than 20 feet (6 meters) deep ¾ to 1 (53 degrees)
Type B	1 to 1 (45 degrees)
Type C	1½ to 1 (34 degrees)

For a complete discussion of excavations, check out OSHA's website (www.osha.gov) under Sloping and Benching–1926 Subpart P –Appendix B.

ENVIRONMENTAL ISSUES

Soil erosion, control of tracking mud onto city or county roads, and prevention of silting up existing storm sewer systems are all critical issues that must be considered before that first shovel full of dirt is taken. Erosion control measures such as silt fencing and hay bale protection around connected storm sewer catch basins are often required by the local building department before any site inspections can take place. Failure to install or properly maintain soil erosion protection can be cause for the local building department to shut the job down, something that no builder wants to happen.

Entry and egress to the site should be protected by a stabilized construction entrance consisting of an area as wide as the entry driveway into the site and possibly 50 feet (165 meters) long. This area will be covered with a layer of 2- to 3-inch (5.08cm to 7.62cm) aggregate 8 to 10 inches deep (20.32cm to 25.4cm). Its purpose is to allow trucks to travel on a stone roadbed a sufficient distance to shed any mud or silt that collected in their tire treads or under their fenders while operating on site. When the stone gets clogged with mud, clay, or silt, it should be turned under and new, clean material added on top. The neighbors and the building inspector will give you high marks.

Silt fence installation and maintenance are a rather simple process and not too expensive; controlling siltation during dewatering operations may not be so simple and inexpensive. Although many builders have been observed pumping out their foundations after a heavy rainfall and allowing reddish brown water to flow onto the street or into storm sewer inlets, better not let any building official see that—you would likely be directed to flush out a few hundred or a few thousand feet of storm sewer downstream from your project or pay a hefty fine or both.

Sediment control can be created by merely pumping water from excavations into a settling basin, allowing the water to disperse and filter down into the soil. Some building departments require a more sophisticated method of sediment control using a sump pit (Figure 9.3) that effectively prevents silt from being discharged with the water. Water from dewatering operations is pumped into the stone pit, where it will be pumped out through a filter; wrapped, perforated, or slotted corrugated metal pipe, or PVC pipe. This discharge water will be relatively clean and can go directly into the existing storm sewer system.

SELECTING EXCAVATING EQUIPMENT

Excavators

Bucket size, boom length, and wheeled or tracked options are factors to consider when renting an excavator, not to be confused with a backhoe loader. Excavators range in power from 54 to 700 horsepower, and boom sizes have reaches up to 50 feet (15 meters), digging depth of 27 feet (8.1 meters), and bucket capacities from ½ cubic yard (.38 cubic meters) to 8½ cubic yards (6.46 cubic meters).

Mini-excavators allow a contractor to work efficiently in confined spaces. These compact excavators with a Japanese heritage are about 5 feet (1.5 meters) wide and 15 feet (4.5 meters) long (including boom), weigh about 5800 pounds (2610 kg) (as compared to the 4500 pounds (2025 kg) of a typical SUV), and have bucket sizes and capacities to fit almost any tight spot. Kobelco, one of the more popular brands, has a 4½ foot wide (1.36 meters), 14.3 hp (SAE net) model that has a bucket with heaped capacity as small as 1.57 cubic feet (.04 cubic meters). There are also compact excavators with hydraulic arms and bucket offset to either the left or right side of the carriage that give the operator a better feel when working in areas requiring close tolerances. At the opposite end of the scale, large excavators, such as

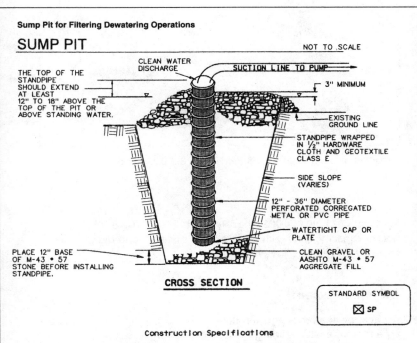

Sump Pit for Filtering Dewatering Operations

SUMP PIT

NOT TO SCALE

CLEAN WATER DISCHARGE

SUCTION LINE TO PUMP

THE TOP OF THE STANDPIPE SHOULD EXTEND AT LEAST 12" TO 18" ABOVE THE TOP OF THE PIT OR ABOVE STANDING WATER.

3" MINIMUM

EXISTING GROUND LINE

STANDPIPE WRAPPED IN ½" HARDWARE CLOTH AND GEOTEXTILE CLASS E

SIDE SLOPE (VARIES)

12" – 36" DIAMETER PERFORATED CORREGATED METAL OR PVC PIPE

WATERTIGHT CAP OR PLATE

PLACE 12" BASE OF M-43 • 57 STONE BEFORE INSTALLING STANDPIPE.

CLEAN GRAVEL OR AASHTO M-43 • 57 AGGREGATE FILL

CROSS SECTION

STANDARD SYMBOL

☒ SP

Construction Specifications

1. Pit dimensions are variable, with the minimum diameter being 2 times the standpipe diameter.

2. The standpipe should be constructed by perforating a 12" to 24" diameter corrugated or PVC pipe. Then wrapping with ½" hardware cloth and Geotextile Class E. The perforations shall be ½" x 6" slits or 1" diameter holes.

3. A base of filter material consisting of clean gravel or #57 stone should be placed in the pit to a depth of 12". After installing the standpipe, the pit surrounding the standpipe should then be backfilled with the same filter material.

4. The standpipe should extend 12" to 18" above the lip of the pit or the riser crest elevation (basin dewatering only) and the filter material should extend 3" minimum above the anticipated standing water elevation.

FIGURE 9.3 Sump pit for sedimentation control.

FIGURE 9.4 Komatsu's 611 HP model with 8.5 cubic yard bucket.

Komatsu's 611 HP PC100SE, has an 8.5 cubic yard (6.46 cubic meters) bucket capacity, can reach 46 feet (13.9 meters), and dig to a depth of nearly 26 feet (7.8 meters) (Figure 9.4).

Other Equipment

A backhoe loader is the familiar rubber-tire John Deere tractor-type machine with loading bucket on one end and digging bucket on the other.

Although bucket attachments are the most common implements attached to wheel loaders such as the Cat 300 series or Komatsu's S6D series, pallet forks, plows, material handling arms, or log and lumber forks are available. Wheel loaders have bucket capacities ranging from 1.5 to 17 cubic yards (1.14 cubic meters to 12.9 cubic meters)

Track loaders don't have the mobility of wheel loaders, their go-anywhere ability and their stability make them candidates for tough, heavy-duty work

The GradAll™ is the premier example of the telescopic handler. Its boom telescopes, while maintaining the ability to rotate with a wrist action that creates a motion that can excavate, backfill, and grade.

The bulldozer blade is not only standard equipment on the Cat D series dozers, for example, but smaller dozer blades are also fitted on some excavators, making them sort of dual-purpose machines.

Bobcat nearly dominated the skid-steerer market until other manu-facturers realized the versatility of and demand for this small machine. Called skid steerers because of the way in which they turn, attach-ments such as general-purpose buckets, augur, backhoe, sweeper, dozer blade, and hoe ram make the skid steerer the equipment of choice for all-around use on small jobs.

SURVEYING TIPS

Layout for site work requires accurate line and grade work. Gener-ally, a surveying firm may provide the initial site layout—staking out property lines, building location, roadways, and so forth. At some point a fast read on existing versus proposed elevation or a check on the invert elevation of some underground utility may be required.

Although lasers seem to be the instrument of choice on large construction sites, the optical instrument has not faded into the past. Levels or transits are still very much in evidence on construc-tion sites, and a brief refresher course on how to use an optical instrument might be of value.

COMPACTION OF SOILS FOR STRUCTURES AND PAVING

Compaction of soil is necessary to support the structures and loads to be placed upon them, to reduce call-backs to re-grade areas where trenching has occurred, and to ensure that sub-base material required for concrete or asphalt paving will stand up to the wear and tear to which it will be subjected.

Proper compaction of soil is not a hit-or-miss proposition but is dependent upon soil composition, moisture content in the soil, lay-ering the soil to be compacted, and compaction equipment specif-ically engineered for that purpose.

What is soil compaction? It is the process of applying energy to loose soil in order to consolidate it, thereby removing all voids and resulting in increased density and increased load-bearing capabili-ties. Not only does compaction increase the soil's load-bearing capacity, but it prevents future settlement, reduces the swelling and contraction that occurs in uncompacted soil, and reduces water seepage intrusion so that any required drainage can be controlled.

fastfacts

You can compact pea gravel! Pea gravel is compactable because the stones are irregular, making them subject to settling if compaction is not performed.

There are several ways to compact soil:

1. Static Force: Rolling over the area to be compacted with a piece of heavy equipment, such as a truck or dozer, so that the weight of the equipment will squeeze the soil particles together. Figure 9.5 represents a typical static force compaction device, a roller with high intrinsic weight.

2. Impact Force: Running a ramming shoe-type machine over the ground so that the soil is actually "kneaded." In fact, one of the most popular compactors, made by the Wacker Corporation, is called a Rammer. Figure 9.6 is a cut-away section of a Rammer (also called a jumping jack due to its up and down motion when compacting) Figure 9.7 is a variation on the jumping jack using vibratory motion.

3. Vibration: The application of a high-frequency vibrating machine over the area to be compacted. Vibratory compactors can be of the walk-behind flat plate type (Figure 9.8) or the riding-type vibratory roller (Figure 9.9)

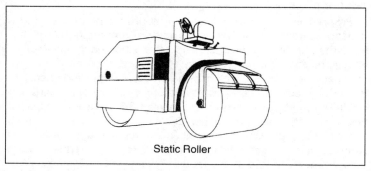

Static Roller

FIGURE 9.5   Typical static force compaction device.

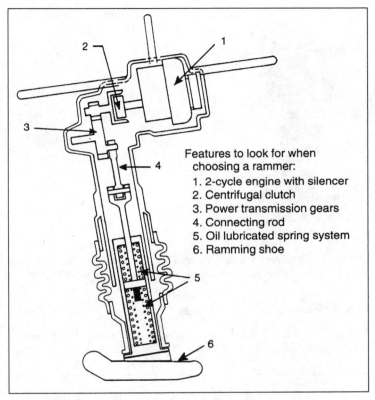

Features to look for when choosing a rammer:
1. 2-cycle engine with silencer
2. Centrifugal clutch
3. Power transmission gears
4. Connecting rod
5. Oil lubricated spring system
6. Ramming shoe

FIGURE 9.6 Cut-away section of a rammer.

The vibration machine is used most effectively when granular soils are to be compacted. The vibration action reduces the friction forces and allows the particles to freely fall under their own weight. As the soil particles are set in motion and they turn and twist, this creates a settling action, filling the air voids in the soil and further increasing compaction.

The impact force machine is the preferred method when cohesive soils such as clay and silt are encountered. The rather flat shape of clay particles does not lend itself to compaction by vibration, but the impact force squeezes out any air pockets, pushes any water to the surface, and presses the particles close together.

Large static rollers such as the one illustrated above are also man-

FIGURE 9.7 Jumping jack rammer with vibratory.

BS 600 Vibratory Rammer

ufactured with vibratory action, and the combination of these two forces are very effective in compacted a wide range of soil types.

Proctor Test

The most frequently used test to determine or verify the degree of compaction achieved is called a Proctor test. A field technician

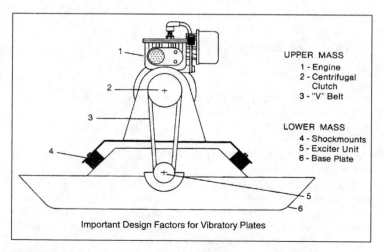

FIGURE 9.8   Schematic of walk-behind flat plate vibratory compactor.

FIGURE 9.9   Riding type vibratory roller.

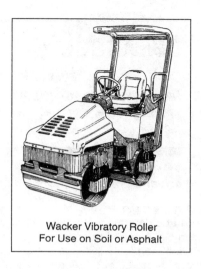

Wacker Vibratory Roller
For Use on Soil or Asphalt

takes a sample of soil and compacts it in a cylinder measuring 4 inches (10.16 cm) in diameter and about 4½ inches (11.4 cm) high. It is filled in three layers, and each one is rammed with a 5.5-pound weight (2.47 kg) raised to a height of 12 inches (30cm) and dropped 25 times on each of three successive layers . The soil is then weighed and then weighed again after the soil is dried in an oven. The difference between "wet" weight (as it came from the excavation) and dried weight (as it came out of the oven) represents the weight of water in the soil, and now the density of the soil can be expressed in terms of pounds per cubic foot. The amount of water in the sample is expressed as a percentage of the dry weight of the soil.

Various amounts of water are added to the soil, and this entire process of filling the container, ramming, weighing, and drying is repeated over and over, and a moisture density curve is created by plotting dry density and percent moisture obtained during these tests.

By selecting a certain point on the curve, it can be determined that at specified moisture levels the soil reaches a maximum density after a specific amount of compaction energy is applied. Subsequent tests confirm moisture levels and density and degree of compaction.

fastfacts

Quick Compactor Selection
Granular soils—use vibratory plate or roller compactor.
Cohesive soils—use rammer or cleated vibratory trench roller.
Mixed soils—use any rammer or trench roller.

Nuclear Density Method

A more rapid method of confirming compaction or the lack thereof can be determined by the nuclear density method. This verification process is based upon the theory that dense soil absorbs more radiation than loose soil. The portable device, about the size of a boom box, is placed directly on the soil to be tested. A rod-shaped device is hammered into the soil and gamma rays from the unit's radioactive source penetrate the soil. The detection of air voids is determined by the number of rays reflected back to the surface of the ground, which are registered on a counter. The counter visually registers the soil density in terms of pound per cubic feet.

Compaction Equipment Maintenance

Dust is the enemy of every compactor, and maintenance of the air filter is critical. Erratic operation in a rammer-type compactor is probably caused by clogged outlet ports at the engine exhaust. Correct by cleaning the ports and inspecting the muffler. Too much oil in the fuel mixture will cause carbon build up and foul the plugs. Use a ratio of 50:1 oil to gas.

Vibratory plates are basically maintenance-free but, once again, air filters need to be checked and the oil in the oil-bath-type air filter changed every 8 hours.

Rollers that are hydraulically driven need a regular change of oil and dirt, and other contaminants must be kept out of the system.

SITE UTILITIES

Rare is the site where some form of underground utility is not required, whether it be installing a single sanitary sewer line to a street connection, extending a storm sewer line into a subdivision or industrial park, or installing underground electrical cabling.

Underground utility construction involves the installation of conduits or pipes of a variety of sizes, shapes, and materials. The basic materials of conduit construction are either precast concrete, iron or steel, thermoplastics, or lightweight aluminum.

Installation of these pipes or conduits has several elements in common. (Note: The terms "pipe" and "conduit" are really interchangeable—both are politically correct!)

1. Excavation and pipe laying: Depending upon the type of soil encountered, the width and depth of the excavatation will be accomplished by either "open cut" or "trench cut" (utilizing sheet piling) or by the use of a trench box (to avoid collapse of the walls of the excavation).

2. Bedding material: Depending upon the type of pipe being installed and the type of soil in the trench, specific "bedding" materials may be required to ensure that the pipe will remain relatively stable in the trench before and during backfill.

3. Conduit selection: Flow rates and materials compatible with the utility being piped will determine the size and material of construction for the conduit selected.

Excavation and Pipe Laying

Open-cut excavation is a rather simple affair, and if room permits, is usually the fastest and most economical method of installing underground pipe, but in some instances trenching may be required in a limited space. And when it occurs, OSHA standards for width and angle of repose for trenching as well as good construction practices may dictate sheathing, shoring, or a trench box.

Shoring of trenches is accomplished by bracing one bank against another; the structural members that transfer the load between trench sides are referred to as *struts*.

Sheathing is defined as the wood planks placed against the trench walls to retain the vertical banks. The horizontal members of the bracing system that form the framework for the sheathing are termed *walers* or *stringers,* and the vertical members are called *strongbacks*.

The four most common sheathing methods are shown in Figure 9.10:

• Open or skeleton sheathing: Consists of a continuous frame with vertical sheathing planks placed at intervals along the open

Types of Sheathing and Shoring

Accurate evaluation of all these factors is usually not possible, so the design and application of temporary bracing systems varies considerably. However, certain methods, materials, and terminology have evolved for stabilizing open trenches and serve as a general guide.

Shoring for trenches is accomplished by bracing one bank against the other, and the structural members which transfer the load between the trench sides are termed **struts.** Wood planks placed against the trench walls to retain the vertical banks are termed **sheathing.** The horizontal members of the bracing system which form the framework for the sheathing are termed **walers** or **stringers,** and the vertical members are termed **strongbacks.**

The four most common sheathing methods are:

- Open or skeleton sheathing
- Close sheathing
- Tight sheathing
- Trench shields or boxes

FIGURE 9.10 Four common sheathing methods *(continued on facing page).*

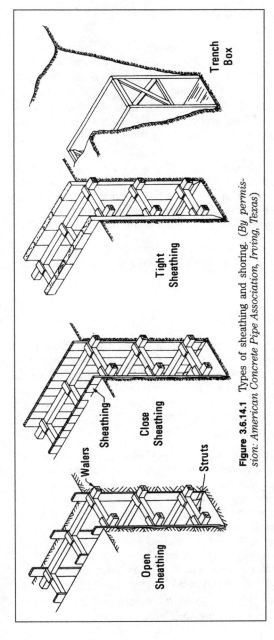

Figure 3.6.14.1 Types of sheathing and shoring. (*By permission: American Concrete Pipe Association, Irving, Texas*)

FIGURE 9.10 Four common sheathing methods (*continued*).

trench; this method of sheathing is used for cohesive, stable soils where groundwater is not a problem

- Close sheathing: Also uses a continuous frame but the vertical sheathing planks are placed side by side so as to form a continuous "retaining wall " effect; this method is used when non-cohesive or unstable soils are encountered

- Tight sheathing: Similar to close sheathing except that the vertical sheathing planks are interlocked; this method is used when saturated soils are encountered; steel sheet piling is sometimes used in lieu of wood

- Trench shields or trench boxes: These heavily braced steel or sometimes wood boxes are inserted into the excavation and moved along as excavation and pipe laying progresses; when trench shields are employed care must be taken to avoid pulling the pipe apart by the friction created in the trench by the move

Note: When renting a trench shield, several factors will influence your selection. Figure 9.11 is a checklist that will assist in making the proper choice.

Conduit and Pipe Selection

Precast concrete pipe; ductile iron pipe, both lined and unlined; thermoplastic pipe (polyvinyl chloride or PVC); chlorinated polyvinyl (CPVC); flexible polyethylene-coated spiral steel tubing; and corrugated metal pipe, either steel, aluminum, or aluminized steel, are the predominant types of underground pipe materials. Each one has a specific purpose and function and requires slightly different joining and installation methods:

1. Precast Concrete pipe: this pipe is manufactured in five common shapes: circular, horizontal elliptical, vertical elliptical, arch, and rectangular. Sizes 4-inch (100 mm) to 36-inch (915 mm) are generally non-reinforced and produced with tongue and groove joints or bell and spigot joints that are sealed with cement mortar or push-on-type rubber gaskets. Sizes ranging up to 36 inches are used for storm sewer applications and are not reinforced but pipe sizes exceeding 3 feet may require reinforcement. Figure 9.12 includes readily available sizes of both reinforced and non-reinforced concrete pipe.

2. Ductile iron and cast iron pipe: The material of choice for water and fire lines in either unlined or cement-lined ver-

Determining Trench Shield Size

If the company does not own a trench box, but plans to rent one, certain data, shown below, must be given to the rental company to ensure that the proper size box is ordered to fit the job at hand.

To size a trench shield

Depth of cut   _____

Soil Conditions*
Type A (25#)   _____
Type B (45#)   _____
Type C (60#)   _____
Hydrostatic   _____

Outside pipe diameter   _____
(Shield 12 in wider than pipe OD)

Pipe length   _____
(Shield 2 to 4 ft longer)

Bucket width   _____
(Inside shield: 12 in less than shield)
(Outside shield: 4 in more than shield)

Machine lift capacity   _____
(1.5 times shield weight at 20-ft radius at grade)

* Soil conditions refer to OSHA classifications. (See Sec. 1.1.5.2 for a full explanation of Type A, B, and C soils.)

FIGURE 9.11 Checklist when renting a trench shield.

sion. This pipe can also be used for storm sewers but it is not cost-effective when compared with concrete or thermoplastic pipe. Ductile iron pipe requires more installation time if shorter than stock lengths have to be cut and fit in the field. Cement-lined ductile iron is frequently used for portable water lines, and polyethylene-coated ductile iron pipe is used when the soils may be corrosive. Ductile iron

Reinforced Concrete Pipe Specificaitons
ASTMC14, ASTMC76 (Bell and Spigot)

ASTM C 14—Nonreinforced Sewer and Culvert Pipe, Bell and Spigot Joint.						
	CLASS 1		CLASS 2		CLASS 3	
Internal Diameter, inches	Minimum Wall Thickness, Inches	Approx. Weight, pounds per foot	Minimum Wall Thickness, Inches	Approx. Weight, pounds per foot	Minimum Wall Thickness, inches	Approx. Weight, pounds per foot
4	5⁄8	9.5	3⁄4	13	7⁄8	15
6	5⁄8	17	3⁄4	20	1	24
8	3⁄4	27	7⁄8	31	1 1⁄8	36
10	7⁄8	37	1	42	1 1⁄4	50
12	1	50	1 3⁄8	68	1 3⁄4	90
15	1 1⁄4	80	1 5⁄8	100	1 7⁄8	120
18	1 1⁄2	110	2	160	2 1⁄4	170
21	1 3⁄4	160	2 1⁄4	210	2 3⁄4	260
24	2 1⁄8	200	3	320	3 3⁄8	350
27	3 1⁄4	390	3 3⁄4	450	3 3⁄4	450
30	3 1⁄2	450	4 1⁄4	540	4 1⁄4	540
33	3 3⁄4	520	4 1⁄2	620	4 1⁄2	620
36	4	580	4 3⁄4	700	4 3⁄4	700

ASTM C 76—Reinforced Concrete Culvert, Storm Drain and Sewer Pipe, Bell and Spigot Joint.				
	WALL A		WALL B	
Internal Diameter, inches	Minimum Wall Thickness Inches	Approximate Weight, pounds per foot	Minimum Wall Thickness, inches	Approximate Weight, pounds per foot
12	1 3⁄4	90	2	110
15	1 7⁄8	120	2 1⁄4	150
18	2	160	2 1⁄2	200
21	2 1⁄4	210	2 3⁄4	260
24	2 1⁄2	270	3	330
27	2 5⁄8	310	3 1⁄4	390
30	2 3⁄4	360	3 1⁄2	450

FIGURE 9.12   Nonreinforced and reinforced concrete pipe dimensions.

pipe is assembled with flange-type or push-on fittings with rubber gaskets or with a lock rig assembly.

3. Thermoplastics: Polyvinyl chloride (PVC) is the most common form of thermoplastic pipe used for storm and sanitary sewer installations and is also used as electrical conduit in many places. PVC electrical conduit is frequently encased in concrete, particularly when it is installed under public walks and road-ways and when it serves as pipe containing primary electrical

cables. Figure 9.13 lists Schedule 40 and Schedule 80 PVC socket weld pipe I sizes from ¼ inch (6.35mm) to 8 inches (20.32 cm).

4. Flexible tubing: Black polyethylene fabric covering a spiral-wound wire framework is often used for foundation drains, French or trench drains, yard drain connections, temporary dewatering operations, and other less demanding roles. Available in rolls with various diameters up to 12 inches (30.48cm), this product is inexpensive to purchase and inexpensive to install.

5. Corrugated metal pipe: Corrugated metal pipe made of galvanized steel, with or without an asphalt coating, is mainly associated with culvert work where large-diameter, lightweight pipe is required to move large volumes of water.

Bedding Materials

Proper bedding of pipe requires some understanding of pipe zone terminology:

1. Foundation: The area directly under the pipe may not require any foundation if the trench bottom is stable and will support a rigid pipe without causing deviation in grade or such flexing of the pipe that may result in failure. If conditions will not permit a structurally stable base, compacted sand or stone is often used for that purpose.

2. Bedding: The material required to bring the trench bottom up to grade and provide uniform longitudinal support. Sand is often used for this purpose.

fastfacts

Expansion of thermoplastic pipe must be considered during installation. PVC pipe will actually increase 2.338 inches in length per 50 feet of pipe at a temperature rise of 70 degrees F (21 degrees C).

PVC Schedule 40/80 Socket Dimensions

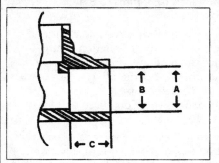

PVC IPS Schedule 40/80 Socket Dimensions

SIZE	PIPE O.D.	ENTRANCE (A)		BOTTOM (B)		MAX. OUT OF ROUND	SCHEDULE 40 SOCKET DEPTH (C) (MIN.)	SCHEDULE 80 SOCKET DEPTH (C) (MIN.)
		MAX.	MIN.	MAX.	MIN.			
1/4	.540	.556	.548	.540	.532	.016	.500	.625
3/8	.675	.691	.683	.675	.667	.016	.594	.750
1/2	.840	.852	.844	.840	.832	.016	.688	.875
3/4	1.050	1.062	1.054	1.050	1.042	.020	.719	1.000
1	1.315	1.330	1.320	1.315	1.305	.020	.875	1.125
1 1/4	1.660	1.675	1.665	1.660	1.650	.024	.938	1.250
1 1/2	1.900	1.918	1.906	1.900	1.888	.024	1.094	1.375
2	2.375	2.393	2.381	2.375	2.363	.024	1.156	1.500
2 1/2	2.875	2.896	2.882	2.875	2.861	.030	1.750	1.750
3	3.500	3.524	3.508	3.500	3.484	.030	1.875	1.875
3 1/2	4.000	4.024	4.008	4.000	3.984	.030	2.000	
4	4.500	4.527	4.509	4.500	4.482	.030	2.000	2.250
5	5.563	5.593	5.573	5.563	5.543	.060	3.000	
6	6.625	6.658	6.636	6.625	6.603	.060	3.000	3.000
8	8.625	8.670	8.640	8.625	8.595	.090	4.000	4.000

FIGURE 9.13   Schedule 40 and Schedule 80 PVC pipe sizes.

3. Haunching: The material used in a zone referred to as "haunching", (the process of placing fill halfway up both sides of the pipe) will supply structural support for the pipe and prevent it from flexing or deflecting or exhibiting joint misalignment when the trench is backfilled and compacted.

4. Initial backfill: The material placed 6 to 12 inches (15.24cm to 30.48cm) above the spring line to provide additional side support; most coming from compaction of the soil in the haunching area.

Figure 9.14 illustrates the various terminology referred to above. There are several classes of bedding materials:

- Class I: Angular stone graded from ¼ inch (6.4 mm) to ½ inch (12.7 mm) including crushed stone, crushed shells, and cinders
- Class II: Coarse sand and gravel with a maximum particle size of 1½ inch (38.1 mm) including various graded sands and gravel containing small percentages of fines; soil types GW, SP, SM, and C (refer back to the Uniform Soil Classification tables) are acceptable.
- Class III: Fine sand and clay gravel including fine sand, sand-clay mixtures, and gravel-clay mixes; soil types GM, GC, SM, and SC are included in this class.
- Class IV: Silt, silty clays (including organic clays), and silts of medium to high plasticity and liquid limits; soil types MH, ML, CH, and CL are included in this class.
- Class V: Soils not recommended for bedding, haunching, or initial backfill; consisting of organic silts, organic clays, and peat, along with other highly organic materials

Positioning the Pipe in the Trench

Batter boards are often used to establish the correct invert elevation and longitudinal location of the underground pipe. Figure 9.15 illustrates how to set up a batter board and locate the position of the pipe and invert (flow line) by the use of grade stakes and grade rod. A plumb bob can also be used to establish the centerline of the pipe or conduit, and, of course, lasers and optical instruments are often the method of choice when installing significant amounts of underground pipe.

Concrete Pipe. The first consideration in laying concrete pipe is the trench width. Sufficient room must be made to not only install the pipe and allow for alignment, but workable room for the installer must also be considered. Bell and spigot joint installation is the most common method of laying concrete pipe.

Thermoplastic Pipe. Thermoplastic pipe joining relies on solvent-welded or -threaded connections. Solvent-welding procedures,

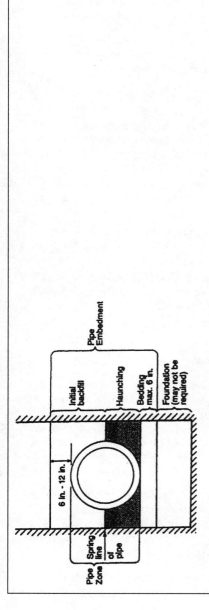

- *Foundation*   Might not be required if the trench bottom is stable and will support a rigid pipe without causing deviation in grade or such flexing of the pipe that will create flexural failure.

- *Bedding*   This material is required to bring the trench bottom up to grade and to provide uniform longitudinal support. Sand is often used for this purpose.

- *Haunching*   This material used in this zone will supply structural support for the pipe and prevent it from deflecting (if it is a flexible pipe) or having joint misalignment when further backfilling and compaction above occurs.

- *Initial backfill*   Material placed 6 to 12 inches above the spring line will only provide additional side support, most of the support coming from compaction of the soil in the haunching area.

FIGURE 9.14   Illustration of pipe bedding terminology.

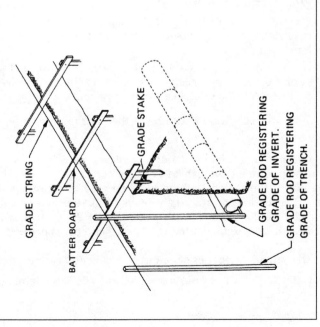

Site Work and Site Utilities   **101**

Setting Up Batter Boards and Determining Trench Widths and Depths

The figure shows how to set up a batter boards so as to establish a center line of underground pipe.

GRADE STRING

BATTER BOARD

GRADE STAKE

GRADE ROD REGISTERING GRADE OF INVERT.

GRADE ROD REGISTERING GRADE OF TRENCH.

FIGURE 9.15   Batter board set-up to position pipe in a trench.

when performed in either hot or cold temperatures, require some special treatment.

Directing Equipment Operators. In case you need to direct an equipment operator during trenching or pipe laying operations, being familiar with standard hand signals for boom equipment operators will come in handy and the illustrations in Figure 9.16 will serve you well.

Site Utility Appurtenances. Manholes are frequently installed whenever there is a change in direction of a sanitary or storm sewer line or in the case of a storm line with an open grate top, where it is required to collect surface water. Precast concrete structures (Figure 9.17) have long replaced the old brick and mortar manholes, and the value of a precast structure is that it can be ordered in as many sections as necessary to attain the height or depth required and installed quickly. The sections of a precast concrete manhole are:

- Base section: Either precast or cast-in-place concrete
- Riser section: Generally available in diameters from 48 to 72 inches (1.2 to 1.8 meters) or larger and in 12-, 24-, or 36-inch (30, 61, or 91 cm) height increments
- Eccentric cone section: Sits on top of the round riser and reduces the diameter down to receive a grade rig and manhole frame and cover
- Grade ring: The cone-shaped precast section that makes the transition from the large diameter manhole to the smaller diameter, standard size of a typical cast iron frame and cover

Storm-sewer manholes can be of the flat top or curb type configuration.

Castings for Sanitary and Storm Manholes. Gray iron castings are corrosion-resistant and exhibit excellent compressive strength. Ductile iron is the material of choice when gray iron castings do not have the load-bearing capacity required for the job. Ductile iron castings are used where loads greater than H20 are required, such as areas subject to forklift traffic, airports, and container ports.

The various categories of cast iron/ductile iron castings are:

- Light duty: Castings recommended for sidewalks, terraces, very light traffic areas

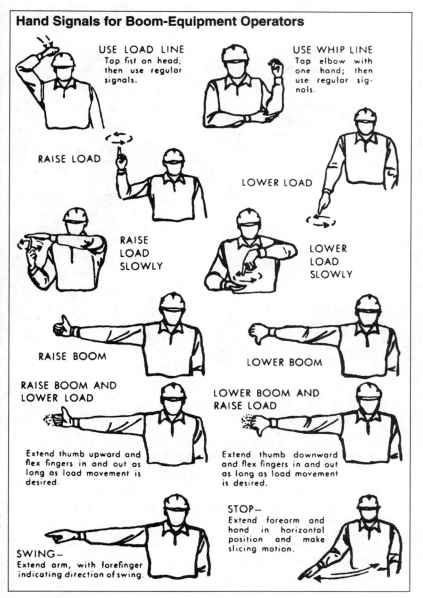

FIGURE 9.16 Hand signals for boom equipment operators.

Storm and Sanitary Manhole Schematics with Sections

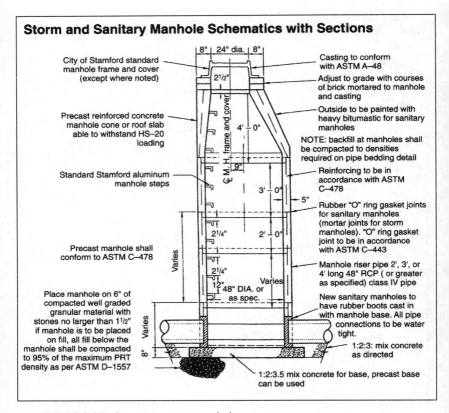

City of Stamford standard manhole frame and cover (except where noted)

Precast reinforced concrete manhole cone or roof slab able to withstand HS–20 loading

Standard Stamford aluminum manhole steps

Precast manhole shall conform to ASTM C–478

Place manhole on 6" of compacted well graded granular material with stones no larger than 1½" if manhole is to be placed on fill, all below the manhole shall be compacted to 95% of the maximum PRT density as per ASTM D–1557

Casting to conform with ASTM A–48

Adjust to grade with courses of brick mortared to manhole and casting

Outside to be painted with heavy bitumastic for sanitary manholes

NOTE: backfill at manholes shall be compacted to densities required on pipe bedding detail

Reinforcing to be in accordance with ASTM C–478

Rubber "O" ring gasket joints for sanitary manholes (mortar joints for storm manholes). "O" ring gasket joint to be in accordance with ASTM C–443

Manhole riser pipe 2', 3', or 4' long 48" RCP (or greater as specified) class IV pipe

New sanitary manholes to have rubber boots cast in with manhole base. All pipe connections to be water tight.

1:2:3: mix concrete as directed

1:2:3.5 mix concrete for base, precast base can be used

FIGURE 9.17 Precast concrete manholes.

- Medium duty: Driveways, parks, ramps, and other installations where wheel loads will not exceed 2000 pounds
- Heavy duty: Castings in this category are suitable for highway traffic or H20 wheel loads of 16,000 pounds minimum
- Extra heavy duty: Airport and port container locations where uniformly distributed loads of 100 to 225 psi will be experienced

When determining the load capacity of a casting, provide the vendor with the following information:

- Total wheel load
- Tire of wheel contact
- Tire pressures

Manhole steps of either steel or aluminum are usually installed between the riser sections to permit easy access to portions of the structure. These foot/hand holds are available in a number of different configurations. Storm sewer frames and grates are available in rectangular shape or round configuration.

Transformer Pads and Enclosures. Local utility companies have their own requirements for transformer pads and clearances required for access. When incoming primary electric service connects to a transformer servicing the building, generally the underground incoming service conduits are encased in concrete, creating a duct bank.

Electric Splice Boxes and Manholes. Depending upon the nature of the incoming electrical service, the local utility company may require the builder to install a splice box similar to Figure 9.18. Another splice-box configuration is shown in Figure 9.19, and a typical frame and grate for these structures are shown in Figure 9.20.

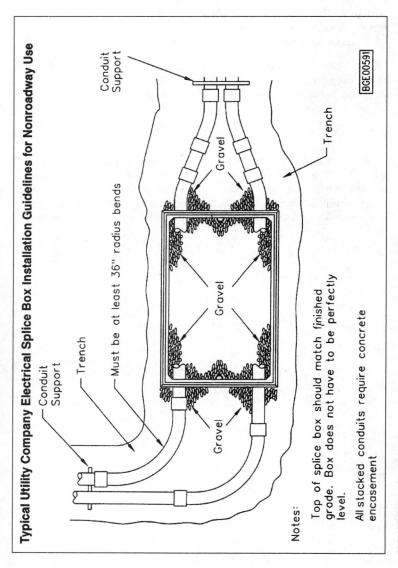

FIGURE 9.18   Typical splice box structure.

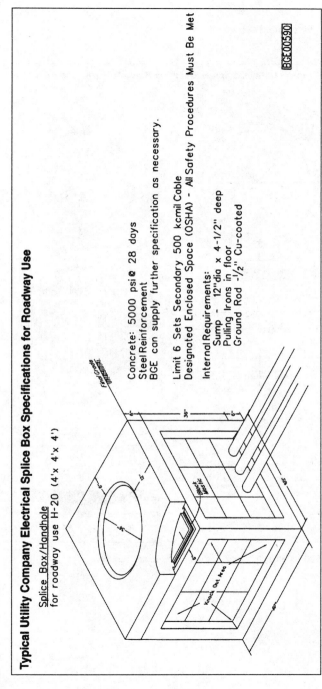

FIGURE 9.19 Pull/splice box—another configuration.

107

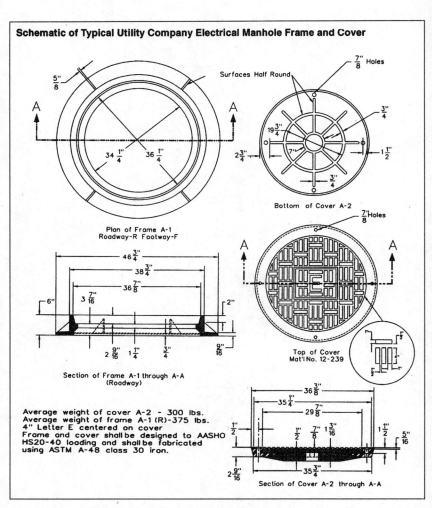

FIGURE 9.20   Pull/splice box cast iron frame and cover.

10

CONCRETE

Concrete, a material first used during the Roman Empire and known then as *concretus,* has often been misunderstood over the past 2500 years.

Concrete is a mixture of 5-7% Portland cement, 14-21% water and 60-75% aggregate. The way in which it is mixed, worked, and consolidated while being placed and allowed to harden can create a durable, long lasting material, or one that fails the test of time and fails to meet construction standards. A successful batch of concrete needs only four elements:

- Precise measurement of water content (very important)
- Type, size, and amount of cement and aggregate
- Type, size, and proper placement of any required reinforcement within the pour

fastfacts

Did you know that adding one gallon of water to a properly designed cubic yard of 3000 psi concrete will cut its compressive strength by as much as 200psi and will increase by about 10%?

- Proper curing procedures during normal, hot, and cold placement

  We will address all four elements in this chapter.

PORTLAND CEMENT

There are many different types of cement available to meet special purpose projects and job conditions. Although Type I and Type III cement are the most commonly used in ready-mix concrete, there are other types that you may want to use when the occasion arises.

- Type I—a general purpose cement used for pavements, floors, reinforced concrete buildings, precast concrete products
- Type II—used where moderate sulfate attack may take place and in structures of considerable mass–large piers, heavy abutments. This cement generates less heat during hydration and at a slower rate than if Type I is used.
- Type III—known as "high early" because it provides higher strength at an earlier stage of curing.
- Type IV—used in concrete mixes where the heat of hydration must be minimized. As a result, this cement type develops strength over a longer period of time. This cement is used in massive concrete pours such as dams.
- Type V—used in concrete mixtures that will be subjected to severe sulfate action such as placement in soils or areas where groundwater contains a high sulfate content.

White Portland Cement - similar in content to Type I or Type III cement, but the manufacturing process is controlled so that the finished product will be white rather than the standard gray color associated with cement.

fastfacts

Did you know?
Portland Cement is not a brand name; it is a generic name for the type of cement used in all types of concrete.

BLENDED HYDRAULIC CEMENTS

Other by-product materials added to and blended with Portland cement have been developed with an eye to energy conservation and include such blends as:

- Type IS—granulated blast furnace slag added to Portland cement
- Type IP—created by adding pozzolan with Portland cement clinker or Portland blast furnace slag cement to pozzolan
- Type I (PM) —pozzolan modified Portland cement
- Type S—a blend of ground granulated blast furnace slag and Portland cement or combination cement, slag, and hydrated lime

    Note: The term "pozzolan" refers to a siliceous or aluminous material having no cementitious value but can chemically react with other properties of cement to form concrete.

MIXING

Freshly mixed concrete should be plastic, a term meaning pliable and capable of being molded. When properly mixed, each grain of sand, cement, and aggregate is encased and held in suspension. During shipment from the batch plant to the construction site, the constituents of the concrete mixture should not separate and should remain in a plastic state while being unloaded and placed at the site. The degree of plasticity can be controlled, not by the addition of water, but by the addition of admixtures known as superplasticizers.

Water and Its Impact on the Quality of Concrete

Reducing water in the concrete mix has many advantages:

- It increases compressive and flexural strength.
- It increases weatherability.
- Creates less volume change.
- Reduced shrinkage crack tendencies.

- Increased bond between successive layers and increased bond with reinforcing steel or wire mesh.
- Lower permeability to water penetration.

The Workability Factor in Concrete

The ease with which concrete is placed and finished is called workability. Slump is a term used to measure the consistency of concrete but some supervisors consider slump and workability as one and the same; low slump = low workability = high slump high workability. Although a low slump (thick consistency) may make for difficult placement, increased workability achieved by adding water to increase slump is not the answer. Increased water may lead to *bleeding,* a condition where water migrates to the top surface of the freshly placed concrete as cement, sand and aggregate settle to the bottom of the mix. One result of excessive bleeding is to create a weak top layer that lacks durability. Entrained air provides some increase in workability and reduces the tendency of the mix to bleed. (More on air entrainment later).

Why Vibration is Used in Concrete Placement

When ready mix concrete is placed, vibration reduces the friction between its particles and makes the mixture more fluid. Stiff mixtures benefit the most by vibrating but when concrete is workable enough (having a much wetter consistency) to consolidate by hand rodding, vibration may not be required.

Heat of Hydration and the Hardening Process

Portland cement is not a simple compound; 90% of its weight is composed of tricalcium silicate, dicalcium silicate, tricalcium aluminate, and tetracalcium aluminoferrite. The chemical action between these compounds and water is called "hydration". The most important chemical reaction between cement and water occurs when a calcium silicate hydrate gel is formed and creates the setting and hardening strength and dimensional stability of the concrete mixture.

When concrete sets, its gross volume stays almost the same, but the amount of pores filled with water and air do not contribute to the strength of the mix; only the paste and aggregate create the strength. At times more water than is necessary for complete hydration is inadvertently added to the mixture, but reduction in water content to minimum levels is essential in creating a strong product.

Heat created as cement progresses through the hydration process can be helpful in certain instances and a problem in others. During cold weather pours, the heat of hydration helps in preventing freezing of the mix, but when massive pours occur, heat of hydration may cause undue stress during the hardening-cooling process.

BACK TO BASICS—THE FOUR ELEMENTS OF A SUCCESSFUL BATCH OF CONCRETE

1. Precise measurement of water

   The water to cement ratio: the weight of the water divided by the weight of cement: Permeability as well as the ultimate strength of the concrete is dependent upon the proper water-cement ratio.

   Proper ratios when exposure to the elements is a concern:

   Concrete exposed to fresh water–Max. water-cement ratio-0.50

   Concrete exposed to freeze-thaw–Max. water-cement ratio-0.45 to 0.50

   For corrosion protection for reinforced concrete–Max. water-cement ratio 0.40

   Proper ratios between water-cement and compressive strength of concrete

   2000 psi at 28 days - 0.82 Non air-entrained 0.74 Air entrained

   3000 psi at 28 days - 0.68 Non air-entrained 0.59 Air entrained

   4000 psi at 28 days - 0.57 Non air entrained 0.48 Air entrained

   5000 psi at 28 days - 0.48 Non air entrained 0.40 Air entrained

2. Type, size and amount of cement and aggregate

   As we have noted previously, there are several types of cement available for general and specialized uses. Cement alone, in a concrete mix would be not only expensive, but the differential shrinkage between internal and external portions of the pour would be unacceptable. Aggregates bound together with the cement/water paste create a concrete mix of high strength with minimal shrinkage and affordable cost.

   The type and size of aggregate, both fine and course, is another important element in the concrete matrix. Fine and course aggregates make up nearly 60% to 75% of the concrete's volume and 70% to 85% of its weight. Fine aggregates are generally sand or crushed stone with particles less than 0.2 inches. Coarse aggregates consist of gravel, crushed quarry rock, boulders, or cobbles. Often recycled concrete is used but it must contain only clean, hard, durable portions of the recycled material. Concrete produced with these types of aggregates will normally weigh between 135 to 160 pounds per cubic foot. Structural lightweight concrete contains aggregates such as rotary kiln expanded clays, shales, slates, palletized or extruded fly ash, or expanded slag and ranges in weight from 35 pounds to 70 pounds per cubic foot as opposed to normal weight aggregates weighing 75 to 110 pounds per cubic foot.

3. Type size and proper placement of any required reinforcement

   Concrete, while extremely strong in compression, is relatively weak in tension.

   In order to compensate for this weakness, reinforcment in various forms and shapes is added to concrete. Reinforcing bars, welded wire mesh, or fiberglass is generally embedded in concrete to add strength and control cracking.

   For reinforcement to perform properly, it must be correctly placed. In footings or slabs, placement directly on grade is a no-no. Keeping the rebar up off the ground with bar supports, in various shapes, sizes, and heights, available from rebar suppliers, is the best way to go. But short of that, a brick support for footing rebar placement is satisfactory. Depending upon the thickness of slab, something less than 2 inches in height is preferable.

   When rebars are installed in vertical walls, they must be secured with tie wire, not to add strength to the bars, but

fastfacts

Before that slab pour, did you remember to pull up that wire mesh so that is doesn't remain right on the vapor barrier?

merely to secure them so that they remain in place during the pour. They are to be placed so that they avoid contact with the forms and remain so during the pour

Reinforcing bar sizes and weights in both U.S. and Metric and ASTM grades are shown in Figure 10.1. Wire Mesh sizes and weight charts,by the more familiar 4 x4 ,6 x 6 and 12 x 12 and also by "W" designation are shown in Figure 10.2.

4. Curing procedures during normal, hot and cold weather

Concrete gains strength as long as any unhydrated cement remains in the mix, provided that the concrete remains moist. When relative humidity within the concrete mixture falls below 80%, or when the temperature of the pour drops below freezing, hydration and therefore strength gain ceases. That is the primary rationale behind the moist curing techniques specified by architects and engineers. Concrete must retain enough moisture during the curing period to allow all of the cement to hydrate.

Normal curing can be accomplished in a number of ways:

<u>Water curing:</u> spraying or misting water on the curing concrete or placing moist burlap over the pour- and keeping it moist are two of the most common methods of applying a water cure.

fastfacts

When using these "membrane" type blankets, inspect them before and during application to insure that there are no large rips or tears that will allow moisture to escape.

BAR SIZE DESIGNATION	WEIGHT POUNDS PER FOOT	NOMINAL DIMENSIONS—ROUND SECTIONS		
		DIAMETER INCHES	CROSS-SECTIONAL AREA-SQ INCHES	PERIMETER INCHES
#3	.376	.375	.11	1.178
#4	.668	.500	.20	1.571
#5	1.043	.625	.31	1.963
#6	1.502	.750	.44	2.356
#7	2.044	.875	.60	2.749
#8	2.670	1.000	.79	3.142
#9	3.400	1.128	1.00	3.544
#10	4.303	1.270	1.27	3.990
#11	5.313	1.410	1.56	4.430
#14	7.650	1.693	2.25	5.320
#18	13.600	2.257	4.00	7.090

Concrete reinforcing bar size/weight chart.

ASTM Standards, Including Soft Metric

Soft metric size	Nom diam mm	Area mm²	Weight factors		Imperial size	Nom diam inches	Area in²	Weight factors	
			kg/m	kg/ft				lb/ft	lb/m
10	9.5	71	.560	.171	3	.375	.11	.376	1.234
13	12.7	129	.994	.303	4	.500	.20	.668	2.192
16	15.9	199	1.552	.473	5	.625	.31	1.043	3.422
19	19.1	284	2.235	.681	6	.750	.44	1.502	4.928
22	22.2	387	3.042	.927	7	.875	.60	2.044	6.706
25	25.4	510	3.973	1.211	8	1.000	.79	2.670	8.760
29	28.7	645	5.060	1.542	9	1.128	1.00	3.400	11.155
32	32.3	819	6.404	1.952	10	1.270	1.27	4.303	14.117
36	35.8	1006	7.907	2.410	11	1.410	1.56	5.313	17.431
43	43.0	1452	11.384	3.470	14	1.693	2.25	7.650	25.098
57	57.3	2581	20.239	6.169	18	2.257	4.00	13.600	44.619

Comparison of Steel Grades

Soft metric			Imperial		
Grade	mPa	psi	Grade	mPa	psi
300	300	43,511	40	257.79	40,000
420	420	60,716	60	413.69	60,000
520	520	75,420	75	517.11	75,000

(*By permission, Concrete Reinforcing Steel Institute, Schramsburg, Illinois*)

FIGURE 10.1 Reinforcing bar sizes/weights-U.S. and metric.

Impervious sheet membrane curing: sealing in moisture by placing an impervious cover such as polyethylene, or a kraft paper composite with a bitumastic adhesive over the pour is another common curing procedure.

Liquid membrane curing: there are a number of different spray-on or roll-on liquid membrane compounds available in the marketplace that act effectively in the curing process.

CURE TIMES

Cure time is a function of time, temperature, and type of cement used in the concrete mix.

The following cure times take these three factors into account.

At 50 degrees F (10 degrees C)-Measured in "days" required

Percentage design strength required	Type cement used in the mix		
	Type I	Type II	Type III
50%	6	9	3
65%	11	14	5
85%	21	28	16
95%	29	35	26

At 70 degrees F (21 degrees C) - Measured in "days" required

Percentage design strength required	Type cement used in the mix		
	Type I	Type II	Type III
50%	6	9	3
65%	11	14	5
85%	21	28	16
95%	29	35	26

fastfacts

A vapor barrier under a concrete slab does provide curing and does not contribute to a weakened bottom The top section of the slab cures faster than the bottom surface but both areas will harden, albeit at different rates.

Welded Wire Fabric (WWF)

| Wire size number | | Nominal diameter, in. | Nominal weight, lb/ft | Area per width (in.²/ft) for various spacings (in) | | | | | | |
Plain	Deformed			2	3	4	6	8	12	16
W45	D45	0.757	1.53	2.70	1.80	1.35	0.90	0.68	0.45	0.34
W31	D31	0.628	1.05	1.86	1.24	0.93	0.62	0.47	0.31	0.23
W20	D20	0.505	0.680	1.2	0.80	0.60	0.40	0.30	0.20	0.15
W18	D18	0.479	0.612	1.1	0.72	0.54	0.36	0.27	0.18	0.14
W16	D16	0.451	0.544	0.96	0.64	0.48	0.32	0.24	0.16	0.12
W14	D14	0.422	0.476	0.84	0.56	0.42	0.28	0.21	0.14	0.11
W12	D12	0.391	0.408	0.72	0.48	0.36	0.24	0.18	0.12	0.09
W11	D11	0.374	0.374	0.66	0.44	0.33	0.22	0.17	0.11	0.08
W10.5		0.366	0.357	0.63	0.42	0.32	0.21	0.16	0.11	0.08
W10	D10	0.357	0.340	0.60	0.40	0.30	0.20	0.15	0.10	0.08
W9.5		0.348	0.323	0.57	0.38	0.29	0.19	0.14	0.095	0.07

FIGURE 10.2 Welded wire mesh sizes.

(continued on next page)

Size	Size									
W9	D9	0.338	0.306	0.54	0.36	0.27	0.18	0.14	0.090	0.07
W8.5	D8	0.329	0.289	0.51	0.34	0.26	0.17	0.13	0.085	0.06
W8		0.319	0.272	0.48	0.32	0.24	0.16	0.12	0.080	0.06
W7.5	D7	0.309	0.255	0.45	0.30	0.23	0.15	0.11	0.075	0.06
W7		0.299	0.238	0.42	0.28	0.21	0.14	0.11	0.070	0.05
W6.5		0.288	0.221	0.39	0.26	0.20	0.13	0.097	0.065	0.05
W6	D6	0.276	0.204	0.36	0.24	0.18	0.12	0.090	0.060	0.05
W5.5	D5	0.265	0.187	0.33	0.22	0.17	0.11	0.082	0.055	0.04
W5		0.252	0.170	0.30	0.20	0.15	0.10	0.075	0.050	0.04
W4.5		0.239	0.153	0.27	0.18	0.14	0.090	0.067	0.045	0.03
W4	D4	0.226	0.136	0.24	0.16	0.12	0.080	0.060	0.040	0.03
W3.5		0.211	0.119	0.21	0.14	0.11	0.070	0.052	0.035	0.03
W3		0.195	0.102	0.18	0.12	0.090	0.060	0.045	0.030	0.02
W2.9		0.192	0.099	0.17	0.12	0.087	0.058	0.043	0.029	0.02
W2.5		0.178	0.085	0.15	0.10	0.075	0.050	0.037	0.025	0.02
W2.1		0.162	0.070	0.13	0.84	0.063	0.042	0.031	0.021	0.02
W2		0.160	0.068	0.12	0.080	0.060	0.040	0.030	0.020	0.02
W1.5		0.138	0.051	0.090	0.060	0.045	0.030	0.022	0.015	0.01
W1.4		0.134	0.048	0.084	0.056	0.042	0.028	0.021	0.014	0.01

(By permission, Concrete Reinforcing Steel Institute, Schramsburg, Illinois)

FIGURE 10.2   FIGURE 10.2   Welded wire mesh sizes. *(continued)*

(continued on next page)

Common Types of Welded Wire Fabric

Style designation (W = Plain, D = Deformed)	Steel area (in ²/ft)		Approximate weight (lb per 100 sq ft)
	Longitudinal	Transverse	
4 x 4-W1.4 x W1.4	0.042	0.042	31
4 x 4-W2.0 x W2.0	0.060	0.060	43
4 x 4-W2.9 x W2.9	0.087	0.087	62
4 x 4-W/D4 x W/D4	0.120	0.120	86
6 x 6-W1.4 x W1.4	0.028	0.028	21
6 x 6-W2.0 x W2.0	0.040	0.040	29
6 x 6-W2.9 x W2.9	0.058	0.058	42
6 x 6-W/D4 x W/D4	0.080	0.080	58
6 x 6-W/D4.7 x W/D4.7	0.094	0.094	68
6 x 6-W/D7.4 x W/D7.4	0.148	0.148	107
6 x 6-W/D7.5 x W/D7.5	0.150	0.150	109
6 x 6-W/D7.8 x W/D7.8	0.156	0.156	113
6 x 6-W/D8 x W/D8	0.160	0.160	116
6 x 6-W/D8.1 x W/D8.1	0.162	0.162	118
6 x 6-W/D8.3 x W/D8.3	0.166	0.166	120
12 x 12-W/D8.3 x W/D8.3	0.083	0.083	63
12 x 12-W/D8.8 x W/D8.8	0.088	0.088	67
12 x 12-W/D9.1 x W/D9.1	0.091	0.091	69
12 x 12-W/D9.4 x W/D9.4	0.094	0.094	71
12 x 12-W/D16 x W/D16	0.160	0.160	121
12 x 12-W/D16.6 x W/D16.6	0.166	0.166	126

*Many styles may be obtained in rolls.

FIGURE 10.2   Welded wire mesh sizes and weights. *(continued)*

Cold Weather Curing

During cold weather, the ready-mix batch plants generally heat the water and may also heat the fine and coarse aggregates. Depending upon geographic location, the period of time when "winter concrete" is shipped will vary. But when announced by the ready supplier, will remain in effect for a specified period during which a slight increase in the cost per cubic yard will be levied. Additional care must be taken during the placement and curing of concrete in cold weather. Temporary protection and temporary heat will be required so that complete hydration, hence curing, takes place in a controlled environment where freezing conditions are absent.

fastfacts

Hot and windy weather is a double whammy when placing concrete, so have your curing process in the ready and don't hesitate to apply it.

Hot Weather Curing

Hot weather placement has its own problems and several conditions may contribute to less-than-desirable results when placing, finishing, and curing concrete is concerned. It starts with the delivery process. Most on-site inspectors will allow concrete to remain in the rotating barrel of the concrete truck for no more than two hours after loading during hot weather. So attention must be paid to the time of loading as indicated on the receiving ticket. Dumping the truck quickly if this two hour margin is approaching is critical. If the truck arrives at the site after the two-hour time allowance, it may be wise to consider rejecting the load, depending upon weather conditions. If rejection is the choice, call the batch plant immediately and advise them of your decision and the reasons for rejection, the time the truck arrived on site, and the time it left the plant!

Curing during hot weather can be trickier than curing in cold weather and the application of a water cure or an impervious or liquid membrane curing procedure must be quick and thorough in order to prevent or retard the escape of moisture from the pour.

CONCRETE ADMIXTURES

The consistency and hardening properties of ready mixed concrete can be changed by the addition of *admixtures,* chemicals that can increase workability, accelerate or delay curing, entrain air for better weathering capability, or adjust other properties in the mix.

The Romans reportedly used oxblood as an admixture and later research revealed that this proved to be an excellent air entraining agent. Oxblood is pretty hard to come by these days so scientists have developed other types of admixtures.

The most commonly used admixtures are:

- Air entrainment
- Water reducers
- Retarders
- Accelerators
- Superplasticizers (another type of water reducers)

Air Entraining Admixtures and Their Functions

The intentional entrapment of millions of tiny (1 millimeter or larger) bubbles in concrete improves resistance to freezing when exposed to water and deicing chemicals. This resistance to freeze-thaw cycles makes it a standard mixture in concrete exposed to the environment. As water freezes it expands about 9% and with no place for this expansion to go, hydraulic pressure will build up in the concrete causing the cured concrete to flake or peel. The millions of microscopic bubbles in the air-entrained concrete provides areas of expansion as these small voids become filled with ice upon freezing, thereby theoretically relieving hydraulic pressure on the surrounding concrete.

The effects of entrained air on concrete properties are listed below.

Property	Effect
Abrasion	Little effect but increased strength does increase abrasion resistance somewhat
Absorption	Little effect
Bleeding	Significant reduction in bleeding
Bond to steel	Decreased
Compressive Strength	Reduced 2% to 5% per percentage point increase in Air
Freeze-thaw	Significantly improves resistance
Heat of hydration	No significant effect
Scaling	Greatly reduced
Shrinkage during drying	Little effect
Slump	Increases approximately 1 inch per ½ to 1% of air
Unit weight	Decreases with increased air
Workability	Increases with air

Total Amounts of Air Content Percent for Concrete Mixes

Exposure Levels

Nom. Max.Aggregate Size	Severe	Moderate	Mild
⅜ inch	7.5%	6%	4.5%
½ inch	7%	5.5%	4%
¾ inch	6%	5%	3.5%
1 inch	6%	4.5%	3%
1½ inch	5.5%	4.5%	2.5%
2 inch	5%	4%	2%
3 inch	4.5%	3.5%	1.5%

Air entrained concrete has more workability than concrete without any entrainment according to one study performed by the Portland Cement Association.

Water Reducing Admixtures

Recalling that the quality of concrete is improved when the addition of water to the mix is reduced, water reducing admixtures can be used to either reduce the quantity of water required to produce a certain slump or increase slump while reducing water-cement ratios.

While regular water reducing admixtures reduce the water content by 5% to 10%, there are high-range admixtures referred to as "Superplasticizers" that can reduce water content by 12% to 30% and still provide workability. Water reducing admixtures can often create increased cracking during the drying process and some water reducing admixtures can retard the concrete's setting time. Superplasticizers are used when intricate bundles of reinforcing bars are required in a form with restricted space. The somewhat soupy mix of the concrete with this admixture will flow around corners and fill in the spaces between closely bundled rebars.

Retarding Admixtures

The introduction of a retarding admixture into the concrete mix *retards* its setting time. These types of admixtures are frequently used to offset the accelerating effect of the concrete pour during very hot weather or can be used when an exposed aggregate finish is required and delaying of the finishing process is desirable. An alternative to using retarders during hot weather pours can be achieved by requesting the ready-mix company to chill either the mixing water or the aggregate or both, at the batch plant.

Accelerating Admixtures

These types of admixtures are used when it is desirable to accelerate the strength of the concrete mix at an early stage after the pour. Calcium chloride is frequently one type of admixture used for this purpose, but as any superintendent knows, chlorides attack metal over the long haul and therefore are to be avoided. Non-chloride bearing accelerators should be sought out if acceleration is required. The advantages of this admixture are outweighed by its tendency to cause increases in drying shrinkage, potential discoloration of concrete, and corrosion of embedded steel reinforcement.

Other accelerating admixtures contain chemicals such Tri-ethanolamine, sodium thiocyanate, calcium formate and calcium nitrite or nitrate. Consider using Type III "Hi-Early" cement as another way of increasing cure time.

Superplasticizers

A form of water-reducing admixture, superplasticizers when added to low slump concrete, create the same workability as high slump concrete. Similar to water reducing admixtures, superplasticizers are more effective, but at a higher cost. A superplasticizer added to 3-inch slump concrete can produce the same effect as a 9 inch slump; however the increased flowability characteristic is short lived, often no more than 30 to 60 minutes, and this admixture, as opposed to others which are batch plant added, is generally added to the concrete at the jobsite. The flowability characteristic of this admixture is especially helpful when:

1. The pour is relegated to thin section placements
2. There are closely spaced areas of reinforcing steel and the concrete must flow around these congestions.
3. When the Tremie pipe delivery system is used in underwater concrete placement.
4. When concrete is being pumped a long distance and reduced pump pressure is required.

CONCRETE STRENGTH

Tests for compressive strength are the tests most familiar to project superintendents; they are generally expressed in terms of PSI or pounds per square inch when the test cylinders have been properly aged for 28 days. Generally test cylinders are made so that they can be "broken" in 7 days, 14 days, 28 days, and sometimes extra cylinders are created in case a 56-day break is required. Reports of these tests are quickly prepared and sent to the field in order to properly monitor the quality of all future pours. But when are test results of early cylinder breaks in the acceptable range?

According to the Building Code Requirement for Reinforced Concrete as set forth in American Concrete Institute (ACI) Reference 318, if the average of all sets of three consecutive strength tests equal or exceed the specified 28 day strength requirement and if no individual cylinder's test result is more than 500 psi below design strength, the compressive strength of the concrete can be considered satisfactory.

Recommended Slumps for Various Types of Concrete Construction

Unreinforced footings and substructure walls	3
Reinforced foundation walls and footings	3
Slabs	3
Pavements	3
Beams and reinforced walls	4
Columns (building)	4

fastfacts

Improper cylinder preparation and storing of test cylinders will give false readings. Rod the cylinder as it is filled and, if a lot of concrete is going to be pored, get a curing box to store the cylinders.

The Slump Test-What It Is, How It is Performed

Slump, as it refers to concrete, is a measure of consistency. The slump test is a method by which the decrease in height of a molded mass of fresh concrete is measured when released from a "slump cone." This slump cone has a round base measuring 8 inches, a top base of 4 inches, and a height of 12 inches. This slump cone is filled with concrete from the ready mix truck in three layers, each one rodded with a bullet-shaped rod (most people use a length of rebar). When the mold has been filled, the top is struck off flush and the mold is lifted to allow the concrete to be released. The amount by which the concrete settles is referred to as "slump"

VAPOR BARRIERS

A vapor barrier should be positioned under all concrete slabs placed on grade, especially if they are to receive a floor finish. A good quality concrete slab with a minimum thickness of 4 inches will resist water infiltration unless there is significant pressure behind the water. However, a concrete slab on grade is not immune to the passage of water vapor and that is why an impermeable membrane such as polyethylene is required to prevent the passage of water vapor. Floor coverings subsequently installed over a concrete slab on grade will effectively seal in the moisture and ultimately cause the flooring material to buckle, blister or begin to disintegrate if the flooring material is carpet. Polyethylene membrane vapor barriers should be 4 mil to 6 mil thickness, and care must be used in avoiding punctures during installation and insuring that sufficient laps are made to prevent the passage of water vapor.

fastfacts

The average amount of vapor emitted from a curing slab is 3 to 4 pounds of vapor per 1,000 square feet in a 24 hour period. Floor treatments installed over slabs with excessive vapor emission are likely to fail.

FORMWORK

Depending upon the application, concrete formwork can be as simple as a 2 x 12 footing plank or as elaborate as an architectural concrete form with a specialty liner. Plywood, steel, and fiberglass are the preferred materials and structural integrity combined with accuracy of installation are the final ingredients in constructing good formwork.

Plywood remains the material of choice for most builders and just about every type and grade of plywood can be used to form concrete. There is a wide variety of plywood available, however, each suited to a specific purpose.

Plyform® is available in two basic grades: Class I and Class II. Class I has Group 1 facings for strength and stiffness a category that includes beech, birch, Douglas fir, western larch and southern pine. Class II Plyform will have Group 2 softer facing: hemlock, luan, cedar, fir, spruce and some pine species.

Structural I Plyform is the strongest plywood forming material. It can withstand the highest loads across and along the panel. B-B Plyform is a non-overlaid plywood with sanded panels on both sides and treated with a release agent at the mill, a process referred to as "mill oiled". However additional applications of a release agent may be required if the panels are not "fresh." These panels can be ordered with sealed edges, if requested. Five to ten uses of B-B plyform can be expected.

HDO Plyform (HDO refers to High Density Overlay) has a hard opaque surface of a thermo-set resin impregnated and bonded to the entire surface of the plywood. This surface will provide a very smooth concrete finish. HDO Plyform, with reasonable care, can be re-used 20 to 50 times.

MDO Plyform (Medium Density Overlay) should not be confused with regular MDO plywood which is meant to receive a painted surface and is not to be used as a concrete forming material. MDO Plyform is normally factory treated with a release agent but should be treated with an additional coat of release agent before its initial use, and after each subsequent use, in order to preserve its surface a bit longer.

Form Facts

- Clean the face of the form between each use
- Apply release agents as required

- When reinforcing steel is required in the form, check to see that it is located within the form correctly and securely.
- Double check the integrity of the bracing; you don't want any blow-outs.
- Will vibration of the concrete be necessary to insure that the pour is relatively free from air pockets, voids and will flow around all portions of the form and any embedded items? Consolidation can be performed by the use of a vibrator immediately after the concrete is poured or by rodding or working with a shovel, tamper, or even one's foot if the pour is wide enough.

There are several forms manufacturers on the market with forms for specialized uses, replaceable faces, and features to fit almost everyone's needs and pocketbook, although they all serve the same basic purpose. See Figures 10.3 and 10.4.

Choosing and Using Form Release Agents

If the form face sticks to the concrete when the form is stripped, either form or concrete will be damaged. A release agent seals the surface of the form, prevents damage to the surface of the concrete, and prolongs the line of the form. There are two basic types of release agents, barrier and chemical.

Barrier agents create a barrier between the form face and the concrete. Diesel fuel and home heating oil are two common barrier type agents, often with a surfactant or wetting agent added to provide better surface adherence. Environmental restrictions have gradually phased out these types of materials. Chemically active release agents have an active ingredient that combines, chemically, with the calcium ions in fresh concrete.

fastfacts

Never use release agents containing silicone or wax if the concrete surface will subsequently be painted; the paint won't adhere.

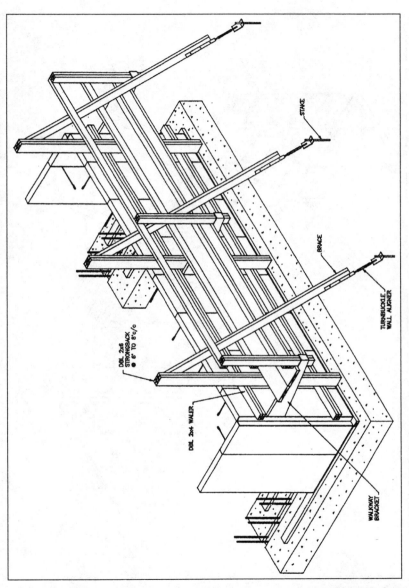

FIGURE 10.3 Typical concrete wall from schematic with walkway bracket installed—one side in place.

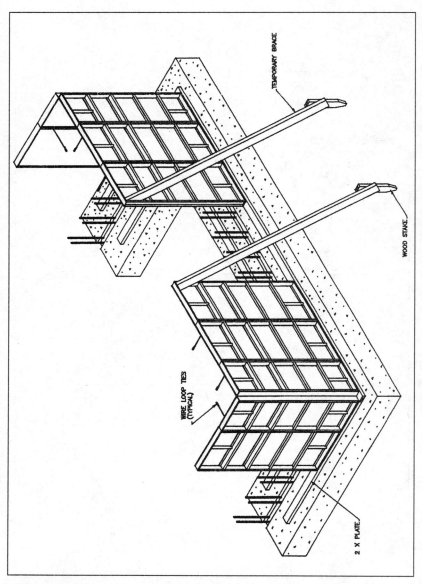

FIGURE 10.4 Typical concrete wall from schematic—one side in place.

- Petroleum oils
- Emulsions- water or oil based
- Nonreactive coatings with volatile solvents
- Chemically active agents
- Waxes

WALL CONCRETE PLACEMENT TIPS

Although vibration is required in order to eliminate voids and fill the form completely, excess over vibration should be avoided since it will tend to separate the aggregate from the mixture. It can also create undue hydraulic pressure that could cause the form to "bow-out". Placement of the reinforcing bars should be checked as the pour continues since pressure on one side of the bars or the other may cause the reinforcement to shift and come dangerously close to the surface of one side of the form. And while the pour continues, assign someone the responsibility of checking on the stability of the forms.

PLACEMENT OF CONCRETE ON METAL DECK

Along with all the other procedures to follow when placing concrete slabs, such as making sure that the wire mesh is up from the bottom, placing concrete on metal decks in hot weather, and particularly when the deck has been exposed to direct sunlight, poses another concern. The deck can be hot enough to cause premature and rapid curing of the concrete; if that is the case, hose down the deck to cool it before the concrete truck arrives. Make sure that excess water is removed from the flutes in the metal deck; broom or screed any excess water to the nearest penetration in the deck.

fastfacts

When using an "elephant trunk" to place concrete in a form, don't let the concrete fall more than 6 feet or else aggregate separation might occur.

Metal decks tend to sag under the added dead load of the concrete, and periodic checking to ensure that the proper depth, and no more, is being placed is essential. Something as simple as a nail driven into a 2 x 4 to mark the proper depth is an easy and quick way to check thickness.

PLACING AND FINISHING THE CONCRETE SLAB

Placement of concrete should start at the farthest point and proceed to the source of the ready-mix truck chute. When placing concrete in vertical forms, overfill slightly and then screen off the excess. Vibrate the rod as required.

DO YOU NEED CONTROL JOINTS?

The term "control joint" is often used synonymously with *construction joint,* but there is a difference between the two. A *control joint* is created to provide for the inevitable movement in the slab as it cures and to induce cracking where it is placed. A *construction joint,* is a bulkhead that is used to end the day's pour. When construction joints are created by bulkheading off a slab pour, steel dowels are usually inserted in the bulkhead to increase load transfer at that point.

SCREEDING OR STRIKING OFF

This initial working of the concrete is to cut off the excess material in a slab pour bringing the top surface to the proper grade. A straightedge is used for this purpose and is worked across the surface of the slab with a sawing motion. This straightedge should have a surplus of concrete along its front edge so that any depressions or low spots ahead will be filled in.

BULLFLOATING

After the straightedge strikes its way across the slab a bullfloat, sometimes referred to as a "darby" is then used to eliminate any remaining high or low spots. This bullfloating must be done before any bleed water accumulates on the surface of the slab

FINAL FINISHING OPERATIONS

When the bleed water has evaporated and the concrete will withstand foot pressure, either power floating or hand floating is necessary to provide a smooth surface. Edging will not only separate the edge form from the slab but will densify the concrete, making it less likely to chip or crack. Floating with either a power trowel or hand trowel will embed aggregate particles just below the surface, remove any slight imperfections in the slab's surface, and compact the surface mortar.

When a smooth hard surface is required, troweling should follow the floating operation. This operation should be delayed until the surface of the concrete has hardened enough so that water and fines will not be brought to the surface. Wait too long and the slab will be difficult to work; do it too soon and scaling, crazing, and dusting will ultimately occur on the finished, cured slab.

BROOMING

The brooming or broom finish on concrete is most often used to create a slip resistant surface. The brooming operation generally follows floating and best results are obtained when a broom made especially for concrete texturing is used. The brooming action should be applied transversely to the direction of traffic.

HARDENED FLOORS

Most often used in industrial or commercial operations to provide added protection against surface abrasion and impact, hardening is not too complicated to be used in residential construction. This

fastfacts

Any finishing operations performed on a slab while bleed water is present can cause major scaling, dusting, or crazing of the slab surface.

fastfacts

If excess water appears on the surface, do not *sprinkle dry cement on the slab; this will cause crazing. Just delay until the water evaporates or is removed with a squeegee.*

hardening process can be obtained by any of the of the following four processes:

1. Densify the slab by hard steel troweling just before the concrete slab reaches its initial set while the slab is becoming hard but has not yet started to hydrate. Hard steel troweling forces entrapped air and moisture out of the surface wear zone forcing the unhydrated cement grains closer together.

2. Increase the density of the concrete mix, maintaining a low water/cement ratio, which will increase compressive strength and, in the process bringing the cement grains closer together

3. Treat the surface with a chemical hardener in the form of soluble silicate or siliconates. Application of these hardeners chemically react with calcium hydroxide, a by-product of the hydration of cement, to produce a hard compound known as calcium silicate hydrate.

4. Apply a shake-on material composed of selected mineral or metal aggregates blended with cement. This material is spread across the newly poured slab and troweled into the surface.

WHY DOES CONCRETE CRACK?

Concrete changes dimensionally as it cures. It can shrink as much as $\frac{1}{16}$th of an inch in ten feet (.4 cm in 3 meters), and depending on the nature of the pour, cracks may appear. Control joints are used to provide a pre-designated spot where the crack will occur and allow the crack to occur in a straight line. Control joints can be created by inserting a type of division strip in the form before pouring or a plastic Zip Strip® after the pour has been placed and troweled. Or the control joint can be saw cut with a diamond blade within 24 hours after the concrete has been placed and finished. Cracks may appear, however, no matter how careful the placing, finishing, and control joint placement or saw cutting.

What's an acceptable crack width? Although opinions vary widely, hairline cracks in concrete slabs or walls are generally acceptable. Many homeowner warranty programs state that:

- Cracks in walls less than ⅛ inch will not be required to be repaired by the builder.

- Cracks in concrete basement floors that exceed 3⁄16 inch in width and ¼ inch in vertical displacement will be repaired by the builder.

- Cracks in concrete garage slabs or structurally attached patio slabs not exceeding ¼ inch in width and less than ¼ inch in vertical displacement are not the responsibility of the builder.

- Separation in concrete stoops that exceed one inch will be repaired by the builder.

The Portland Cement Association lists tolerable crack widths for reinforced concrete as follows:

Exposure Condition	Tolerable crack width (in inches)
Dry air or protective membrane	0.016
Humidity,moist air,soil exposure	0.012
Deicing chemical exposure	0.007
Seawater and seawater spray, wetting/drying	0.006
Water retaining structures	0.004

REPAIRING CRACKS

There are a number of commercially produced crack repair materials on the market today from elastomeric (caulking) materials to epoxys. An elastomeric caulking compound designed for use in concrete can close many hairline cracks.

These cracks should be "chased" first and cleaned out with air pressure, in some cases brushing vigorously with a stiff bristled brush will do. Filling the crack with an acceptable caulking material or epoxy should do the trick.

Small cracks (0.002 inches [0.05 mm] in vertical concrete walls, primarily foundation walls, can be effectively sealed by epoxy injection. Holes about ⅞ inch in diameter are drilled to bridge the crack and are placed 3-5 feet on center. Epoxy is then forced into the holes under pressure to effectively seal the crack. The area around the crack must be completely dry if this method is used, and should be performed when it has been ascertained that the crack is stable and not growing.

To determine crack stability, use a thick pencil like a carpenter's pencil and a straightedge to mark the apex (end) of the crack at both ends. Draw one line horizontally across the crack and one vertically across the crack at each apex. If the crack widens or lengthens, it should be easy to discern by inspecting the pencil marks.

ALLOWABLE TOLERANCES IN CAST-IN-PLACE CONCRETE

The American Concrete Institute (ACI) in their *Manual of Concrete Practice* states that no structures are exactly level, plumb or precisely straight and true; some variations and tolerances are required to provide both designer and contractor with acceptable levels of performance.

With respect to cast-in-place concrete:

Plumb: In lines and surfaces of columns, piers and walls

In any ten vertical feet	¼ inch
Maximum for the total height of the structure	1 inch when structure does not exceed 100 feet

For exposed corner columns and other conspicuous lines

In any 20 vertical feet	¼ inch
Maximum for the total height of the structure	½ inch

When the total height does not exceed 100 feet

Level: From grades and elevations specified

Slab soffits, ceilings, beam soffits, measured before removal of supporting shores

In any ten horizontal feet	+/- ¼ inch
In any bay or in any 20 feet	+/- ⅜ inch
Maximum for total length of structure	+/- ¾ inch

Exposed sills, lintels, parapet, horizontal grooves

In any bay or in 20 feet	+/- ¼ inch
Maximum for total length of structure	+/- ½ inch

Cross sectional dimensions of columns, beams, walls, and slab thicknesses

Up to 12 inches in thickness	+ ⅜ inch–¼ inch
More than 12 inches	+ ½ inch–⅜ inch

Footings
    Horizontal dimensions (formed)   + 2 inches–½ inch
    Horizontal dimensions (unformed)   + 3 inches
To receive masonry construction
    Alignment in 10 feet   +/- ⅓ inch
    Maximum for entire length - 50 feet   +/- ½ inch
    Level in 10 feet   +/- ¼ inch
    Maximum for entire length - 50 feet   +/- ½ inch

Slabs based upon class of tolerances for flatness

Class AA: depressions in slabs between high spots shall be not greater than ⅛ inch below a 10-foot straightedge

Class AX: depressions in slabs between high spots shall be not greater than ³⁄₁₆ inch below a 10 foot straightedge

Class BX: depressions in slabs between high spots shall not be greater than ⁵⁄₁₆ inch below a 10 foot straightedge

Class CX: depressions I slabs between high spots shall not be greater than ½ inch below a 10 foot straightedge

One Cubic Yard of Concrete Will Place

Thickness	Square Feet
2 inches	162
2½ inches	130
3 inches	108
3½ inches	93
4 inches	81
4½ inches	72
5 inches	65
5½ inches	59
6 inches	54
6½ inches	50
7 inches	46
7½ inches	43
8 inches	40
8½ inches	38
9 inches	36
9½ inches	34
10 inches	32.5
10½ inches	31
11 inches	29.5
11½ inches	28
12 inches	27
15 inches	21.5
18 inches	18
24 inches	13.5

fastfacts

One cubic foot of:
   Cement weighs 94 pounds
   Fine aggregate weighs 105 pounds
   Coarse aggregate weighs 100 pounds

Quantities of Concrete for Footings and Walls

Footing Size (")	Cubic Feet of concrete per LF	Cubic Yards of Concrete per 100 LF
6 x 12	0.50	1.9
8 x 12	0.67	2.5
8 x 16	0.89	3.3
10 x 12	0.83	3.1
10 x 16	1.11	4.1
10 x 18	1.25	4.6
12 x 12	1.00	3.7
12 x 24	2.00	7.4

Wall Thickness:

Thickness in inches	Square Feet
3	108
3½	93
4	81
5	65
5½	59
6	54
6½	50
7	46
7½	43
8	40

Adding One Gallon of Water to a Yard of 3000 psi concrete:

   Increases slump by about one inch
   Reduces compressive strength by as much as 200 psi
   Wastes the effect of ¼ bag of cement
   Increases potential for shrinkage by about 10%
   Decreases freeze-thaw resistance by 20%
   Decreases resistance to attack by deicing salts

11

MASONRY

The Egyptians are credited with building the first masonry structure, the Great Pyramid of Giza, in the year 10,000 B.C. The first commercially manufactured brick was made in England in the mid 1600s, and methods of manufacture and installation basically remain the same.

In the 1930s, cavity wall construction was introduced into the United States from Europe as a means of controlling moisture. Through the years more sophisticated masonry units, mortars, and reinforcing materials have been developed, but we still lay up masonry walls one unit at a time.

The three basic components of masonry construction are:

- Masonry units: clay, concrete, and combinations thereof
- Mortar: in various compressive strengths and a wide range of colors
- Reinforcement: horizontal ladur type, vertical type, and various masonry ties

MASONRY UNITS

Bricks are manufactured in a wide variety of colors, textures, and patterns. There are two basic groupings, modular and non-modular, each of which offers a multitude of sizes.

Modular brick sizes

Unit	Dimensions (Inches)			Width-Height-Length W-H-L			Specified Dimensions Joint Thickness (Vertical)	
Modular	4	2⅔	8	3⅜	2¼	7⅝	⅜	3C = 8 inches
				3½	2¼	7½	½	
Engineer Modular	4	3⅓	8	3⅜	2¾	7⅝	⅜	5C = 16 inches
				3½	2¹³⁄₁₆	7½		
Closure Modular	4	4	8	3⅝	3⅝	7⅝	⅜	1C = 4 inches
				3½	3½	7½	½	
Roman	4	2	12	3⅝	1⅝	11⅝		2C = 4 inches
				3½	1½	11½	½	
Norman	4	2⅔	12	3⅝	2¼	11⅝	⅜	3C = 8 inches
				3½	2¼	11½		
Engineer Norman	4	3½	12	3⅝	2¾	11⅝	⅜	5C = 16 inches
				3½	2¹³⁄₁₆	11½	½	
Utility	4	4	12	3⅝	3⅝	11⅝	⅜	1C = 4 inches
				3½	3½	11½	½	

Metric equivalents for modular brick sizes

4 x 2⅔ x 8 = 10.16cm x 5.95cm x 20.32cm
3⅜ x 2¼ x 7⅝ = 8.57cm x 5.63cm x 18.1cm
3½ x 2¼ x 7½ = 8.75cm x 5.63cm x 18.75cm

4 x 3 – ⅓ x 8 = 10.16cm x 8.33cm x 20cm
3⅜ x 2¾ x 7⅝= 8.57cm x 5.63cm x 18.1cm
3½ x 2¹³⁄₁₆ x 7½ = 8.75cm x 5.474cm x 18.75cm

4 x 4 x 8 = 10.16cm x 10.16cm x 20cm
3⅝ x 3⅝ x 7 – ⅝ = 9.2cm x 9.2cm x 18.1cm
3½ x 3½ x 7½ = 8.75cm x 8.75cm x 18.75cm

4 x 2 x 12 = 10.16cm x 5cm x 30cm
3⅝ x 1⅝ x 11⅝ = 9.2cm x 4.127cm x 29.52cm
3½ x 1½ x 11½ = 8.75cm x 3.75cm x 28.75cm

4 x 2⅔ x 12 = 10.16cm x 5.95cm x 30cm
3⅝ x 2¼ x 11 – ⅝ = 9.2cm x 5.63cm x 29.52cm
3½ x 2¼ x 11½ = 8.75cm x 5.63cm x 28.75cm

4 x 3½ x 12 = 10.16cm x 8.75cm x 30cm
3⅝ x 2¾ x 11⅝ = 9.2cm x 5.63cm x 29.52cm
3½ x 2 – ¹³⁄₁₆ x 11½ = 8.75cm x 5.47cm x 28.75cm

4 x 4 x 12 = 10.16cm x 10.16cm x 30cm
3⅝ x 3⅝ x 11⅝ = 9.2cm x 9.2cm x 29.52cm
3½ x 3½ x 11½ = 8.75cm x 8.75cm x 28.75cm

Note: 2C, 5C etc refers to number of courses and "inches" refers to height of that coursing.

Other modular brick sizes				
Nominal Size		Specified Dimensions (Inches)		Joint Thickness	Vertical Coursing
W H L		W H L			
4 6 8		$3\frac{1}{2}$ $5\frac{1}{2}$ $\frac{1}{2}$		$\frac{1}{2}$	2C = 12 inches
4 8 8		$3\frac{1}{2}$ $7\frac{1}{2}$ $7\frac{1}{2}$		$\frac{1}{2}$	1 C = 8 inches
6 $3\frac{1}{2}$ 12		$5\frac{1}{2}$ $2\frac{3}{16}$ $11\frac{1}{2}$		$\frac{1}{2}$	5 C = 16 inches
6 4 12		$5\frac{1}{2}$ $3\frac{1}{2}$ $11\frac{1}{2}$		$\frac{1}{2}$	1C = 4 inches
8 4 12		$7\frac{1}{2}$ $3\frac{1}{2}$ $11\frac{1}{2}$		$\frac{1}{2}$	1 C = 4 inches
8 4 16		$7\frac{1}{2}$ $3\frac{1}{2}$ $15\frac{1}{2}$		$\frac{1}{2}$	1C = 4 inches

Metric equivalents for other modular brick sizes

4 x 6 x 8 = 10.16cm x 15cm x20cm
$3\frac{1}{2}$ x $5\frac{1}{2}$ x $7\frac{1}{2}$ =8.75cm x 13.75cm x 18.75cm

4 x 8 x 8 = 10.16cmx20cmx20cm
$3\frac{1}{2}$ x $7\frac{1}{2}$ x $7\frac{1}{2}$ = 8.75cm x 18.75cm x 18.75cm

6 x 3 1/2 x 12 = 15cm x 8.75cm x 30cm
$5\frac{1}{2}$ x $2\frac{3}{16}$ x $11\frac{1}{2}$ =13.75cm x 5.47cmx28.75cm

6 x 4 x 12 = 15cm x 10.16cm x 30cm
$5\frac{1}{2}$ x $3\frac{1}{2}$ x $11\frac{1}{2}$ = 13.75cm x 8.75cm x 28.75cm

8 x 4 x 12 = 20cm x 10.16cm x 30cm
$7\frac{1}{2}$ x $3\frac{1}{2}$ x $11\frac{1}{2}$ = 18.75cm x 8.75cm x 28.75 cm

8 x 4x 16 = 20cm x 10.16 cm x 40 cm
$7\frac{1}{2}$ x $3\frac{1}{2}$ x $15\frac{1}{2}$ = 18.75 x 8.75cm x 38.75 cm

Other conversions for the previous page; $\frac{1}{2}$" = 1.25cm, 12" = 30.48cm, 4" = 10.16cm, 8" = 20.32cm

Note: Specified dimensions may vary somewhat from manufacturer to manufacturer.

Non-modular brick coursing				
Unit	Specified Dimensions (Inches)		Joint Thickness	Vertical Coursing
	W H L			
Standard	3⅝ 2¼ 8		3/8	3C = 8 inches
	3½ 2¼ 8		½	
Engineer	3⅝ 2¾ 8		⅝	5C = 16 inches
Standard	3½ 2¹³⁄₁₆ 8		½	
Closure	3⅝ 3⅝ 8		3/8	1C = 4 inches
Standard	3½ 3½ 8		½	
King	3 2¾ 9⅝		⅜	5C = 16 inches
	3 3⅝ 9¾		⅜	
Queen	3 2¾ 9¾		⅜	5C= 16 inches

Metric equivalents for non-modular brick coursing	
3⅝ x 2¼ x 8 = 9.20cm x 5.63cm x 20.32 cm	⅜" = 9.5 mm
3½ x 2¼ x 8 = 8.75cm x 5.63cm x 20.32cm	½" = 1.25 cm
3⅝ x 2¾ x 8 = 9.20cm x 6.98cm x 20.32cm	
3½ x 2¹³⁄₁₆ x 8 = 8.75cm x 5.47cm x 20.32cm	
3⅝ x 3⅝ x 8 = 9.20 cm x 9.20cm x 20.32cm	
3½ x 3½ x 8 = 8.75cm x 8.75cm x 20.32	
3 x 2¾ x 9⅝ = 7.5cm x 6.98cm x 24.08cm	
3 x 3⅝ x 9¾ = 7.5cm x 9.20 x 24.37 cm	
3 x 2¾ x 9¾ = 7.5 cm x 6.87cm x 24.37 cm	
3 x 2⅝ x 8⅝ = 7.5cm x 6.58cm x 20.16 cm	

Note: Specified dimensions may vary within this range from manufacturer to manufacturer.

Figure 11.1 illustrates the various types and dimensions of modular and non-modular bricks. Brick positions in the wall are illustrated in Figures 11.2 and 11.3.

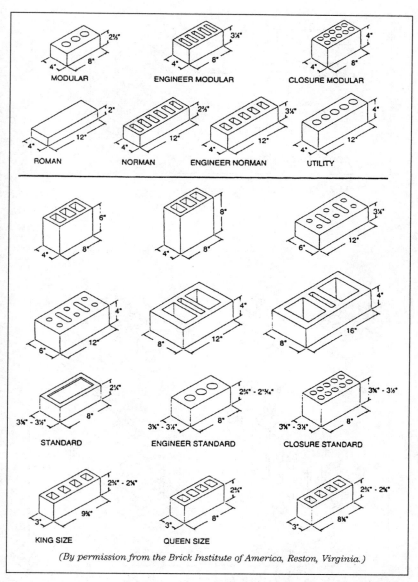

FIGURE 11.1 Modular (above) and nonmodular (below) brick sizes.

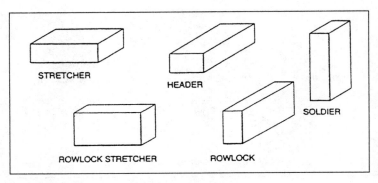

FIGURE 11.2 Brick positions in a wall.

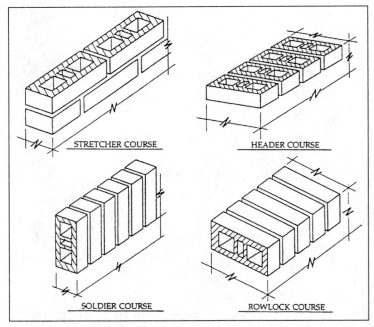

FIGURE 11.3 Brick orientation (illustrated).

TRADITIONAL BOND PATTERNS

The following are some of the most common bond patterns:

- Running bond: the simplest of all brick structures; this pattern consists of all stretchers, and metal wall ties are used when this type of wall is built as a cavity or veneer-type structure.
- Common or American bond: a variation on the running bond, this pattern introduces a course of full-length headers at regular intervals, generally every fifth, sixth, or seventh row.
- English bond: consists of alternate courses of headers and stretchers; the headers are centered on the stretchers, and joints between the stretchers and headers in all courses are aligned vertically.
- English cross or Dutch bond: a variation on the English bond; this pattern differs in that vertical joints between the stretchers in alternate courses do not align vertically.
- Flemish bond: each course of brick consists of alternate stretchers and headers; headers in alternate courses are centered over the stretchers in the intervening courses; half brick or "snapped" headers can be used where structural bonding between the two wythes is not required.
- Block or stacked bond: there is no overlapping of units because all vertical joints are aligned; generally this patterned wall is bonded to the backing with rigid steel ties and reinforcement in the horizontal mortar joints.

CONCRETE MASONRY UNITS (CMUs)

CMUs come in a variety of shapes, sizes, and textures. Their basic constituents are cement, water, and gravel, molded and steam-cured for one to two days. CMUs are not as water-repellent as bricks, but they can be sprayed with a water repellent or a water-repellent admixture can be mixed in with the other ingredients. Basic uses for concrete block, also known as concrete brick, are as backup for veneer walls, cavity wall construction, and single or multi-wythe reinforced walls.

There are two basic grades of CMUs:

- Type N: for use in architectural veneers in exterior walls where high strength and resistance to moisture infiltration is required; compressive strength is 3500 psi.
- Type S: a general grade where moderate or average strength and moisture resistance is required; compressive strength is not less than 2500 psi.

Sizes, Shapes, and Surface Treatments

Architectural CMUs

- Split face: CMU with rough texture on one face, made by producing a double wide unit, say 16 x 16 inches , and splitting it in half to create two 8-x-16-inch "split face" blocks.
- Split face fluted: CMU made in a similar fashion to the previous one, but during the casting process, recesses are created in the

fastfacts

Number of blocks needed to build a wall

Size	No. required per 100 SF
4 x 4 x 16	*225*
10cm x 10cm x 40cm	
6 x 4 x 16	*225*
5cm x 10cm x 40cm	
8 x 4 x 16	*225*
20cm x 10cm x 40cm	
4 x 8 x 16	*112.5*
10cm x 20cm x 40cm	
6 x 8 x 16	*112.5*
15cm x 20cm x 40cm	
8 x 8 x 16	*112.5*
20cm x 20cm x 40cm	
12 x 8 x 16	*112.5*
30cm x 20cm x 40cm	

center of the dual block so that when it is split, flutes appear along the face, one, three, or five depending how many were cast into the mold.

- Scored block vertical grooves: generally ⅜ inch square, are cast into the face of the block; these scores can be vertical, creating the look of two 8-inch blocks in a 16-inch long block, or the score can run horizontally.

- Ground face block: the exposed face of the block is ground to create a smooth appearance.

- Glazed block: a ceramic coating is fired onto the basic CMU; available in almost every color of the rainbow and in many textures.

- Slump block: a CMU resembling an adobe block with an irregular slumped effect on its face.

- Concrete paving units: patio blocks manufactured in a variety of shapes, sizes, colors, and textures; some concrete paving units are used by state highway departments to stabilize soil on bridge embankments and prevent erosion.

Structural CMUs are available in a wide variety of shapes and configurations (Figure 11.4).

STONE

Both native and imported natural stone and man-made stone products are as varied as the shades of the rainbow. Stone used in wall construction falls into three basic categories: rubble (round rocks), flagstone (flat, irregular pieces), and ashlar (dimensioned stone cut into slices for laying in coursed or non-coursed patterns). Some walls use the stone as is while other types, primarily made of flat stones, can be coped to fit as close as pieces in a jigsaw puzzle. Stone can be used either as a veneer, generally supported via masonry ties fastened to concrete or steel-studded or sheathed walls or laid up as a solid stone wall.

Stone materials commonly available include the following:

- Fieldstone: appropriately named, this stone is found in open fields in many parts of the country.

- Granite: a hard, durable stone, of igneous origin, composed of quartz, feldspar, and mica; used in ashlar and rubble walls and

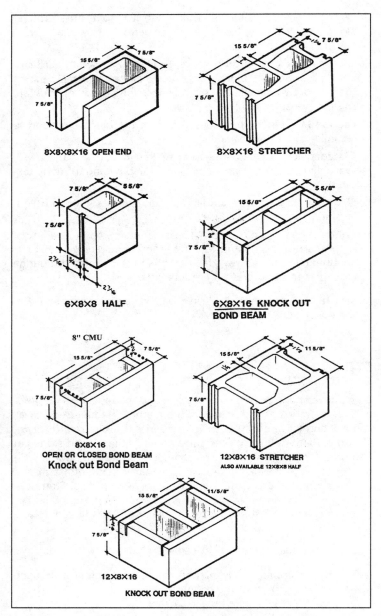

FIGURE 11.4 Wide variety of CMU shapes. *(continued on next two pages)*

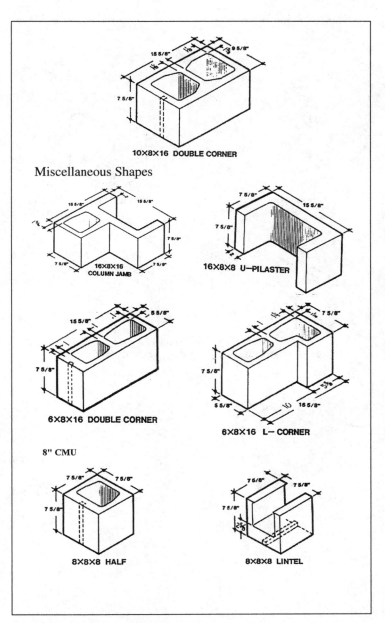

Miscellaneous Shapes

8" CMU

FIGURE 11.4 Wide variety of CMU shapes *(continued).*

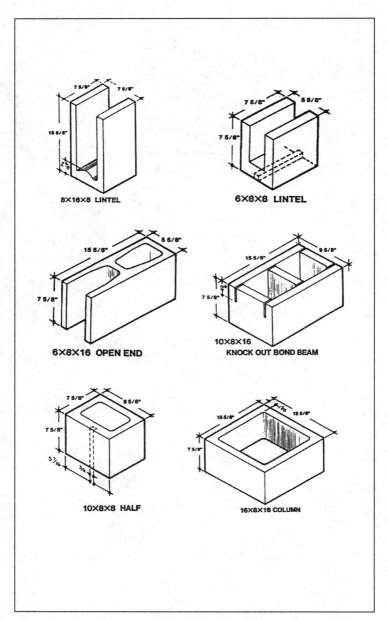

FIGURE 11.4 Wide variety of CMU shapes *(continued)*.

wall panels (where it is often ground and polished), stair treads, copings, sills, and curbing.

- Limestone: a sedimentary rock that is available to three different types:
  1. Oolitic: calcite-cemented rock formed of shells and small aquatic creature shell fragments; it has no cleavage lines and is uniform in composition and color.
  2. Dolomitic: somewhat crystalline in character, available in a greater variety of colors and textures, stronger than oolitic in compression and tensile strength
  3. Crystalline: consists of calcium carbonate crystals, light gray to gray in color, with high compressive and tensile strength; used in wall panels, window sills, copings, and mantles

- Natural lava: flowing from extinct volcanoes, this material may have some rock deposits embedded in its mass; lava is very lightweight, only 20% of the weight of granite; building materials are generally in the shape and form of rubble and used as a veneer.

- Travertine: a sedimentary rock formed on the earth's crust by the evaporation of water from hot springs; most frequently used as an interior finish material, either as flooring or wall panels

- Marble: a metamorphic rock well known by type: Carrara, Vermont, onyx, and brecciated; some marbles are suited for exterior use while others deteriorate when exposed to the weather and should be used for interiors only.

- Schist: a metamorphic rock that splits easily into slabs; made primarily of silica and iron oxides, it comes in a range of colors from white to red to gold to brown, blue, and green; often used for wall panel facings, both interior and exterior, sidewalks and landscaping stones on patios and decks.

- Serpentine: an igneous rock composed largely of magnesium silicate, it is a dense and homogenous rock that is often sliced into thin slabs of ¾ to 1³⁄₁₆ inches thick and used for wall panels, stair treads, window sills, and stair landings.

- Sandstone: a sedimentary rock made of grains of quartz bound together with silica, iron oxide, or clay; hardness is dependent upon the type of "cement"; because sandstone is quite porous (30%), its textured nature lends itself to finish applications.

- Slate: a metamorphic rock that has as its primary characteristic the ability to be separated into thin sheets, anywhere from ¼ inch to 2 to 3 inches in thickness; used as flooring material, ornamental stair railings, and window trim; color varies from gray to black to green and some red/purple shades.

- Quartzite: grains of quartz cemented together with silica form a coarse, crystalline rock; colors range from ivory to brown to red and gray; usually is produced in ashlar form with pieces 2 to 4 inches thick.

ADOBE

Adobe is Spanish for "plaster," which in turn derives from an Arabic word for sun-dried bricks. About 20% of the world's population lives in adobe-type dwellings, and in some places like Santa Fe, New Mexico, adobe homes are very much in vogue.

Adobe is a mixture of clay (25%-40%), sand, and water, with straw often mixed in. Sand acts as a filler, and the clay and water act as binders. Adobe bricks, typically 10 x 4½ x 16 inches and weighing about 35 pounds, can be made by pouring the mixture into forms. The bricks are brittle and don't travel well.

MORTAR

Mortar is the bonding agent that holds, or bonds, all of the masonry units together. Bond strength is the crucial element in mortar mixes; it differs from its near relative, concrete, where compressive strength is the most important physical property.

Mortar serves four functions:

- It bonds the masonry units together and seals the space between them.
- It compensates for dimensional variations in the masonry units so that a high degree of levelness can be attained.
- It bonds to the reinforcing steel in the masonry wall.
- It provides decoration in both texture of tooled joints and color.

Mortar Types

- Type M: high compressive strength (2500 psi average) with greater durability than other types, generally recommended for unreinforced masonry walls below grade

- Type S: reasonably high compressive strength (1800 psi) with great tensile strength; generally recommended for reinforced masonry walls where maximum flexural strength is required
- Type N: mid-range compressive strength (750 psi average); suitable for general above-grade masonry construction for parapets and chimneys
- Type O: low compressive strength (350 psi average); suitable for interior non-load-bearing masonry walls
- Type K: very low compressive strength (75 psi average); occasionally used for interior non-load-bearing masonry walls where permitted by code

Workability or plasticity of the mortar is an essential characteristic in all mortar mixes. Mortar must have cohesive and adhesive qualities when it makes contract with the masonry units. Hardness or high strength is not necessarily a measure of durability. Mortar that is stronger than the masonry units to which it is applied might not "give," thereby causing stress and resulting in cracking of the assembly or spalling of materials.

Mortar Additives

Just like its cement cousin, additives for mortar are used to change its curing rate or install other qualities in the mixture:

- Accelerators: speed up the setting time of mortar by as much as 30% to 40%; *check the chemical composition of the accelerator to ensure that it doesn't contain any chlorides that might attack reinforcing or anchors.*
- Retarders: extend the "life" of mortar by as much as 4 to 5 hours; they slow down the set time of mortar when temperatures exceed 70 degrees F.
- Water repellents: reduce water absorption; particularly important when single wythe walls are being built.
- Bond modifiers: improve adhesion to the block; particularly useful when installing glass block.
- Corrosion inhibitors: used in marine environments where salt air could penetrate the mortar and corrode wall reinforcement.

Mortar Testing

Mortar is tested in a manner similar to concrete, checking for compliance with compressive strength. Mortar samples are collected in what is referred to as "cubes" and tested by the "prism" method. For those interested in learning more about the actual method of testing, check out ASTM E 447, Method B.

Figure 11.5 includes a list of compressive strengths of mortar based upon the type of cement used in its formulation, and Figure 11.6 lists the allowable compressive strength per CMU unit and the allowable compressive strengths and gross cross-sectional areas for Types M, S, and N mortar.

Type of cement	Minimum compressive strength, psi				ASTM designation
	1 day	3 days	7 days	28 days	
Portland cements					C150-85
I	—	1800	2800	4000*	
IA	—	1450	2250	3200*	
II	—	1500	2500	4000*	
	—	1000†	1700†	3200*†	
IIA	—	1200	2000	3200*	
	—	800†	1350†	2560*†	
III	1800	3500	—	—	
IIIA	1450	2800	—	—	
IV	—	—	1000	2500	
V	—	1200	2200	3000	
Blended cements					C595-85
I(SM), IS, I(PM), IP	—	1800	2800	3500	
I(SM)-A, IS-A I(PM)-A, IP-A	—	1450	2250	2800	
IS(MS), IP(MS)	—	1500	2500	3500	
IS-A(MS), IP-A(MS)	—	1200	2000	2800	
S	—	—	600	1500	
SA	—	—	500	1250	
P	—	—	1500	3000	
PA	—	—	1250	2500	
Expansive cement					C845-80
E-1	—	—	2100	3500	
Masonry cements					C91-83a
N	—	—	500	900	
S	—	—	1300	2100	
M	—	—	1800	2900	

*Optional requirement.
†Applicable when the optional heat of hydration or chemical limit on the sum of C2S and C3A is specified.
Note: When low or moderate heat of hydration is specified for blended cements (ASTM C595), the strength requirements is 80% of the value shown.
(By permission from the Masonry Society, ACI, ASCE from their manual Building Code Requirements for Masonry Structures.)

FIGURE 11.5 Compressive strength of mortars made with various types of cement.

	Allowable compressive stresses[1] gross cross-sectional area, psi (MPa)	
Construction; compressive strength of unit, gross area, psi (MPa)	Type M or S mortar	Type N mortar
Solid masonry of brick and other solid units of clay or shale; sand-lime or concrete brick:		
8000 (55.1) or greater	350 (2.4)	300 (2.1)
4500 (31.0)	225 (1.6)	200 (1.4)
2500 (17.2)	160 (1.1)	140 (0.97)
1500 (10.3)	115 (0.79)	100 (0.69)
Grouted masonry, of clay or shale; sand-lime or concrete:		
4500 (31.0) or greater	225 (1.6)	200 (1.4)
2500 (17.2)	160 (1.1)	140 (0.97)
1500 (8.3)	115 (0.79)	100 (0.69)
Solid masonry of solid concrete masonry units:		
3000 (20.7) or greater	225 (1.6)	200 (1.4)
2000 (13.8)	160 (1.1)	140 (0.97)
1200 (8.3)	115 (0.79)	100 (0.69)
Masonry of hollow load bearing units:		
2000 (13.8) or greater	140 (0.97)	120 (0.83)
1500 (10.3)	115 (0.79)	100 (0.69)
1000 (6.9)	75 (0.52)	70 (0.48)
700 (4.8)	60 (0.41)	55 (0.38)
Hollow walls (noncomposite masonry bonded) Solid units:		
2500 (17.2) or greater	160 (1.1)	140 (0.97)
1500 (10.3)	115 (0.79)	100 (0.69)
Hollow units	75 (0.52)	70 (0.48)
Stone ashlar masonry:		
Granite	720 (5.0)	640 (4.4)
Limestone or marble	450 (3.1)	400 (2.8)
Sandstone or cast stone	360 (2.5)	320 (2.2)
Rubble stone masonry Coursed, rough, or random	120 (0.83)	100 (0.69)

FIGURE 11.6 Allowable compressive stresses for masonry.

(continued on next page)

Net area compressive strength of units, psi (MPa)	Moduli of elasticity[1] E, psi × 10⁶ (MPa × 10³)	
	Type N mortar	Type M or S mortar
6000 (41.3) and greater	—	3.5 (24)
5000 (34.5)	2.8 (19)	3.2 (22)
4000 (27.6)	2.6 (18)	2.9 (20)
3000 (20.7)	2.3 (16)	2.5 (17)
2500 (17.2)	2.2 (16)	2.4 (17)
2000 (13.8)	1.8 (12)	2.2 (15)
1500 (10.3)	1.5 (10)	1.6 (11)

[1]Linear interpolation permitted.
(By permission from the Masonry Society, ACI, ASCE from their manual Building Code Requirements for Masonry Structures.)

FIGURE 11.6 Allowable compressive stresses for masonry *(continued).*

MASONRY FLASHINGS

Flashings in masonry construction provide barriers against water infiltration and are used around any openings and sometimes terminations of masonry walls. For example, a masonry opening into which a door or window frame is to be installed will be "flashed" to seal off the head and, in the case of a window, the sill.

Flashing materials are constructed of a special type of woven fabric, paper bonded to a bitumastic material or flexible metal, generally aluminum or copper.

Flashings are also used where masonry construction joins a dissimilar material; for example, when a plywood substrate and shingled roof and a brick chimney meet or when a wood patio roof is attached to a brick wall. In both cases, the joint between brick and wood needs to be protected against water infiltration; the insertion

fastfacts

Dam that window sill! Don't forget to turn up the ends of window-sill flashings to create a dam and prevent water penetration, not allowing it to migrate beyond the end of the sill and into the wall.

of flashing creates a barrier to water penetration while providing a path to divert water away from this critical joint.

Figure 11.7 illustrates a typical masonry veneer wall with either steel stud or CMU backup and the proper installation of flashing, drainable material in the cavity, and weep holes.

WEEPING

Cavity or veneer-type masonry wall construction is designed to create an air gap, or space, between the inside face of the exterior masonry unit and the face of the backup wall, whether it is a stud wall or another type of wall, most typically block. This "air gap" serves two purposes: as an insulating medium where additional insulation can be placed (rigid type) and as a passageway for water that penetrated the outer layer to find its way out of the wall system.

Weeps are the answer. They are small-diameter ⅜" plastic tubing or ropes especially made to act as a wick to draw water out or simply

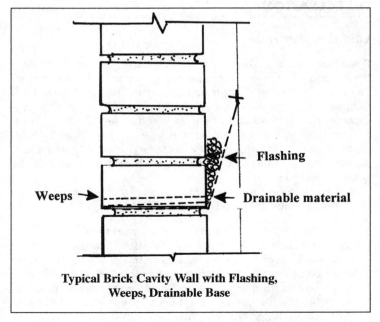

Typical Brick Cavity Wall with Flashing, Weeps, Drainable Base

FIGURE 11.7 Typical flashing detail for cavity wall construction.

the omission of a mortar joint every four or five bricks in the coursing above where the flashing has been tucked into the mortar joint.

In either case, effective weeping will prevent the buildup of water in the cavity, preventing its migration to the brick veneer, increasing efflorescence or to the interior of the partition, causing dampness and the growth of mold.

Keeping the cavity clean of excess mortar droppings and providing a drainage material at the base of the flashing are also essential elements in good cavity wall construction. By placing a 2- to 3-inch layer of some inert, drainable material at the base of the flashing, a repository for any water that penetrates the cavity is created, providing a passageway to the weep.

This area must constantly be checked while the masonry work is being laid up. Too much mortar buttered on the brick or block to receive the next course will fill up this drainable area quickly. Also check thoroughly to avoid damp areas later. When excess mortar droppings occur, they must be carefully removed to avoid puncturing the flashing material below.

INSTALLATION

Cold Weather Tips

Precautions must be taken when endeavoring to install masonry during cold weather, when temperatures remain no higher than 40 degrees F (5 degrees C) or below during the daytime. Even if the temperature in the sun is above 40 degrees F (5 degrees C), in the shade it may be close to 32 degrees F (0 degrees C), so care must be taken during installation as well as after the day's work is done. Drape insulated blankets over the new work or use a tented enclosure in order to contain some temporary heat if temperatures dip below freezing after sundown.

As temperatures reach freezing or below, more of the construction materials must be protected against the cold.

1. Materials delivered to the site need to be protected. Piles of sand must be covered to prevent the water it contains from freezing. Bags of mortar and pallets or stacks of bricks or blocks must be covered with waterproof tarps to prevent any moisture from entering and freezing. When large projects are to be undertaken, it's a good idea to built a temporary shed to keep materials in during cold weather.

2. Water, the one obvious ingredient that needs attention, must be heated, but so must the bricks or blocks if they are exposed to temperatures lower than 20 degrees F (3.5 degrees C). A temporary enclosure with a gas or kerosene heater will "condition" or "climatize" the masonry units. This enclosure could also contain at least a partial batch of sand required for the day's mortar mix.

3. A change in mortar mix may be considered with the substitution of Type III cement (high early) and hydrated lime in lieu of the Type I cement and slaked quicklime.

4. Admixtures in the mortar can help, but care must be taken in their selection. On projects where an architect or engineer is involved in either design or construction inspections, any proposed use of an admixture should be presented for review and comment.

5. Temporary protection is required when temperatures take a sharp dip downward at the end of the day, and during cold weather insulated blankets and materials for temporary enclosures must be at the site or readily available after hours. A few oil or gas-fired heaters, with sufficient fuel, stored on the site will provide extra insurance in case they are needed.

Glass Block

Glass block is often used in exterior as well as interior walls to "open up" an area to outside light and bring brightness into an interior room, stairwell, or, in the case of residential construction, a bathroom or bedroom, while at the same time preserving privacy within the area.

Here's how to install glass block:

1. Cover the sill area with a heavy coat of asphalt emulsion. Allow emulsion to dry at least 2 hours before placing mortar.

2. Adhere expansion strips to jambs and head. Make certain that the expansion strip extends to the sill.

3. Set a full mortar bed joint, applied to the sill.

4. Set the lower course of block. Maintain a uniform joint width of ¼ inch (6.35mm) to ⅜ inch (9.5mm) plus or minus ⅛ inch

(3.175mm). All mortar joints must be full and not furrowed. Steel tools must not be used to tap block into position. (Place a rubber crutch tip on the end of the trowel to tap the block into position). Do not realign, tap, or otherwise move the block after initial placement. For some glass blocks a typical mortar joint is ⅜ inch (9.5mm). It may be necessary to use wedges in the mortar joints of the lower courses to prevent the mortar from being squeezed out.

5. Install panel reinforcing every 16 inches (40.64cm) on center maximum in the horizontal mortar joint and in the joints immediately above and below all openings in the panel. Where panel anchors are used at jambs and heads, in lieu of channels or chase surrounds, install panel anchors in the same joints (16 inches on center maximum) as the panel reinforcing. However, at panel corners, anchors should be placed in each mortar joint, both at the jamb and head, 24 inches (50.8cm) on each side of the corner. Install panel anchors across the head joint spaced 16 inches on center maximum. Run reinforcing continuously from end to end of the panels. Lap reinforcing not less than 6 inches (15.24 cm) whenever it is necessary to use more than one length. Install reinforcing as follows:

    • Place lower half of mortar in bed joint. Do not furrow.

    • Press reinforcing into place.

    • Cover panel reinforcing with upper half of mortar bed and trowel smooth. Do not furrow.

6. Place full mortar bed for joints not requiring panel reinforcing; do not furrow. Maintain uniform joint width.

7. Set succeeding courses of block; the space at head panel and jambs must remain free of mortar for caulking and sealants.

8. Use only wooden or rubber-tipped tools when tapping glass blocks in place.

9. Strike joints while mortar is still plastic and before the final set. Remove surplus mortar from faces of glass blocks and wipe dry. Tool joints smooth and concave before mortar takes the final set. Remove any edges from lower courses at this time and point up voids with mortar. At this time

remove and clean all excess mortar from jamb, head, and other expansion joint locations.

10. After final mortar set (usually 24 hours) install packing tightly between glass panel and jamb and head locations. Leave space for sealants.

11. Apply sealant evenly to the full depth of recesses as indicated on the drawings and in accordance with the manufacturer's recommendations.

To clean the glass block installation:

1. Remove surplus mortar from the cases of the glass block at the time the joints are struck or tooled. Mortar should be removed while it is still plastic, using a clean, wet sponge or a normal household scrub brush with stiff bristles.

2. Do not use harsh cleaners, acids of any strength, abrasives, or alkaline materials when cleaning glass block. Never use steel wool or a wire brush to remove mortar from the face of the glass block.

3. Final mortar removal is accomplished with a clean, wet sponge or cloth. Rinse the sponge or cloth frequently in clean water to remove abrasive particles that could scratch the block. Allow any remaining film on the block to dry to a powder.

4. After all organic materials, caulking, etc., have been applied, remove any excess caulking or sealant materials with a commercial solvent such as xylene, toluene, mineral spirits, or naphtha, followed by a normal wash and rinse. Be careful not to damage caulking materials by using too much solvent during this cleaning operation.

5. The final cleaning of the glass panel or wall should be done when the wall is not exposed to sunlight. Start at the top and wash with generous amounts of clean water. Dry all water from the block and change the cloth frequently to avoid picking up mortar particles that could scratch the glass. To remove dry powder from the glass surfaces, use a clean, dry, soft cloth. For stubborn or hard to remove power or stains, the use of "extra fine" steel wool (grades 000 or 0000) is suggested; however, this type of cleaning should be tried first in an unobtrusive area to ensure that the glass surface is not scratched.

ACCEPTABLE TOLERANCES IN MASONRY CONSTRUCTION

The Masonry Standards Joint Committee in conjunction with the American Concrete Institute developed a list of tolerances for the erection of masonry walls. These tolerances represent acceptable deviations from the specified dimensions indicated on plans or in specifications.

In cross section or elevation	− ¼ inch (6.4mm) to + ½ inch (13mm)
Mortar joint thickness	
Bed	+/− ⅛ inch (3.2 mm)
Head	− ¼ inch (6.4 mm) to ⅜ inch (9.5 mm)
Collar*	− ¼ inch (6.4 mm) to ⅜ inch (9.5 mm)
Variation from level	
Bed joints	+/− ¼ inch (6.4mm) to +/− ½ inch (13mm) per 10 feet maximum
Top surface of bearing walls	+/− ¼ inch (6.4mm) +/− ½ inch (13mm) per 10 feet maximum
Variation from plumb	+/− ¼ inch (6.4mm) to +/− ½ Inch (13mm) in 10 feet +/− ⅜ inch (9.5mm) to ½ inch (13mm) in 20 feet

* vertical space between wythes of masonry that is to be filled with mortar

ESTIMATING COURSINGS IN BRICK AND BLOCK WALLS

When determining the number of courses of brick or block required to create window and door openings, louver openings, and the like, the masonry coursing, both vertical and horizontal, must be determined prior to beginning the wall construction. The following charts for brick and block coursing will come in handy for that purpose. Figure 11.8 includes the nominal height of brick and block walls by coursing. Figure 11.9 includes the nominal length of CMU walls by stretchers and also the nominal height of CMU walls by courses.

GROUTING OF CMU WALLS

Some CMU walls require grouting if they are being constructed in a seismic zone, while others require it to enhance their load-bearing capabilities and still others are grouted in order to produce a tall wall that retains stability. There are two basic types of wall grouting: low-lift grouting, which is grouting of no more than 5 feet in height at one time; and high-lift grouting, which entails grouting a full story height in one day. Both of these operations are not to be performed by masons unfamiliar with grouting procedures.

PREVENTING MASONRY FAILURES

Although masonry walls are extremely durable, "old age" and neglect can take their toll on even the most durable structure. When inspecting a masonry facade for potential problems and restoration, a number of contributing factors must be considered. It is often necessary to cut out a small piece of wall in the area or areas where failure is observed. The ability to look into the cavity, if there is one, may indicate the areas that are failing and the reason for their failure.

The following checklist will help in this investigation process:

1. When initially built, were all ties and anchors installed as required?

2. Were the ties properly installed (i.e., embedded adequately in the bed joint and connected to the back-up correctly)?

3. Does there appear to be excessive differential wall movement caused by thermal expansion, settlement, or freeze and thaw cycles?

4. Were the proper size and type of ties and anchors used to avoid stresses that exceed the facade material's capacity?

5. Were the proper types of expansion and control joints installed at the proper locations?

6. Have the ties, anchors, fasteners, relieving angles, and lintels corroded because of moisture trapped within the wall or cavity?

COURSES	REGULAR 4 2¼" bricks + 4 equal joints =					MODULAR 3 bricks + 3 joints =	CONCRETE BLOCKS	
	10" 1/4" joints	10 1/2" 3/8" joints	11" 1/2" joints	11 1/2" 5/8" joints	12" 3/4" joints	8"	3 5/8" blocks 3/8" joints	7 5/8" blocks 3/8" joints
1	2½"	2⅝"	2¾"	2⅞"	3"	2 11/16"	4"	8"
2	5"	5¼"	5½"	5¾"	6"	5 5/16"	8"	1'4"
3	7½"	7⅞"	8¼"	8⅝"	9"	8"	1'0"	2'0"
4	10"	10½"	11"	11½"	1'0"	10 11/16"	1'4"	2'8"
5	1'0½"	1'1⅛"	1'1¾"	1'2⅜"	1'3"	1'1 5/16"	1'8"	3'4"
6	1'3"	1'3¾"	1'4½"	1'5¼"	1'6"	1'4"	2'0"	4'0"
7	1'5½"	1'6⅜"	1'7¼"	1'8⅛"	1'9"	1'6 11/16"	2'4"	4'8"
8	1'8"	1'9"	1'10"	1'11"	2'0"	1'9 5/16"	2'8"	5'4"
9	1'10½"	1'11⅝"	2'0¾"	2'1⅞"	2'3"	2'0"	3'0"	6'0"
10	2'1"	2'2¼"	2'3½"	2'4¾"	2'6"	2'2 11/16"	3'4"	6'8"
11	2'3½"	2'4⅞"	2'6¼"	2'7⅝"	2'9"	2'5 5/16"	3'8"	7'4"
12	2'6"	2'7½"	2'9"	2'10½"	3'0"	2'8"	4'0"	8'0"
13	2'8½"	2'10⅛"	2'11¾"	3'1⅜"	3'3"	2'10 11/16"	4'4"	8'8"
14	2'11"	3'0¾"	3'2½"	3'4¼"	3'6"	3'1 5/16"	4'8"	9'4"
15	3'1½"	3'3⅜"	3'5¼"	3'7⅛"	3'9"	3'4"	5'0"	10'0"
16	3'4"	3'6"	3'8"	3'10"	4'0"	3'6 11/16"	5'4"	10'8"
17	3'6½"	3'8⅝"	3'10¾"	4'0⅞"	4'3"	3'9 5/16"	5'8"	11'4"
18	3'9"	3'11¼"	4'1½"	4'3¾"	4'6"	4'0"	6'0"	12'0"
19	3'11½"	4'1⅞"	4'4¼"	4'6⅝"	4'9"	4'2 11/16"	6'4"	12'8"
20	4'2"	4'4½"	4'7"	4'9½"	5'0"	4'5 5/16"	6'8"	13'4"
21	4'4½"	4'7⅛"	4'9¾"	5'0⅜"	5'3"	4'8"	7'0"	14'0"
22	4'7"	4'9¾"	5'0½"	5'3¼"	5'6"	4'10 11/16"	7'4"	14'8"
23	4'9½"	5'0⅜"	5'3¼"	5'6⅛"	5'9"	5'1 5/16"	7'8"	15'4"
24	5'0"	5'3"	5'6"	5'9"	6'0"	5'4"	8'0"	16'0"
25	5'2½"	5'5⅝"	5'8¾"	5'11⅞"	6'3"	5'6 11/16"	8'4"	16'8"

26	5' 5"	5' 8 1/4"	5' 11 1/2"	6' 2 3/4"	6' 6"	5' 9 5/16"	8' 8"	17' 4"
27	5' 7 1/2"	5' 10 7/8"	6' 2 1/4"	6' 5 5/8"	6' 9"	6' 0"	9' 0"	18' 0"
28	5' 10"	6' 1 1/2"	6' 5"	6' 8 1/2"	7' 0"	6' 2 11/16"	9' 4"	18' 8"
29	6' 0 1/2"	6' 4 1/8"	6' 7 3/4"	6' 11 3/8"	7' 3"	6' 5 5/16"	9' 8"	19' 4"
30	6' 3"	6' 6 3/4"	6' 10 1/2"	7' 2 1/4"	7' 6"	6' 8"	10' 0"	20' 0"
31	6' 5 1/2"	6' 9 3/8"	7' 1 1/4"	7' 5 1/8"	7' 9"	6' 10 11/16"	10' 4"	20' 8"
32	6' 8"	7' 0"	7' 4"	7' 8"	8' 0"	7' 1 5/16"	10' 8"	21' 4"
33	6' 10 1/2"	7' 2 5/8"	7' 6 3/4"	7' 10 7/8"	8' 3"	7' 4"	11' 0"	22' 0"
34	7' 1"	7' 5 1/4"	7' 9 1/2"	8' 1 3/4"	8' 6"	7' 6 11/16"	11' 4"	22' 8"
35	7' 3 1/2"	7' 7 7/8"	8' 0 1/4"	8' 4 5/8"	8' 9"	7' 9 5/16"	11' 8"	23' 4"
36	7' 6"	7' 10 1/2"	8' 3"	8' 7 1/2"	9' 0"	8' 0"	12' 0"	24' 0"
37	7' 8 1/2"	8' 1 1/8"	8' 5 3/4"	8' 10 3/8"	9' 3"	8' 2 11/16"	12' 4"	24' 8"
38	7' 11"	8' 3 3/4"	8' 8 1/2"	9' 1 1/4"	9' 6"	8' 5 5/16"	12' 8"	25' 4"
39	8' 1 1/2"	8' 6 3/8"	8' 11 1/4"	9' 4 1/8"	9' 9"	8' 8"	13' 0"	26' 0"
40	8' 4"	8' 9"	9' 2"	9' 7"	10' 0"	8' 10 11/16"	13' 4"	26' 8"
41	8' 6 1/2"	8' 11 5/8"	9' 4 3/4"	9' 9 7/8"	10' 3"	9' 1 5/16"	13' 8"	27' 4"
42	8' 9"	9' 2 1/4"	9' 7 1/2"	10' 0 3/4"	10' 6"	9' 4"	14' 0"	28' 0"
43	8' 11 1/2"	9' 4 7/8"	9' 10 1/4"	10' 3 5/8"	10' 9"	9' 6 11/16"	14' 4"	28' 8"
44	9' 2"	9' 7 1/2"	10' 1"	10' 6 1/2"	11' 0"	9' 9 5/16"	14' 8"	29' 4"
45	9' 4 1/2"	9' 10 1/8"	10' 3 3/4"	10' 9 3/8"	11' 3"	10' 0"	15' 0"	30' 0"
46	9' 7"	10' 0 3/4"	10' 6 1/2"	11' 0 1/4"	11' 6"	10' 2 11/16"	15' 4"	30' 8"
47	9' 9 1/2"	10' 3 3/8"	10' 9 1/4"	11' 3 1/8"	11' 9"	10' 5 5/16"	15' 8"	31' 4"
48	10' 0"	10' 6"	11' 0"	11' 6"	12' 0"	10' 8"	16' 0"	32' 0"
49	10' 2 1/2"	10' 8 5/8"	11' 2 3/4"	11' 8 7/8"	12' 3"	10' 10 11/16"	16' 4"	32' 8"
50	10' 5"	10' 11 1/4"	11' 5 1/2"	11' 11 3/4"	12' 6"	11' 1 5/16"	16' 8"	33' 4"

FIGURE 11.8 Nominal height of brick and block walls by coursing.

Estimating Concrete Masonry

NOMINAL LENGTH OF CONCRETE MASONRY WALLS BY STRETCHERS
(Based on units 15⅝" long and half units 7⅝" long with ⅜" thick head joints)

LENGTH OF WALL	NO. OF UNITS	LENGTH OF WALL	NO. OF UNITS	LENGTH OF WALL	NO. OF UNITS	LENGTH OF WALL	NO. OF UNITS	LENGTH OF WALL	NO. OF UNITS	LENGTH OF WALL	NO. OF UNITS
0'-8"	½	20'-8"	15½	40'-8"	30½	60'-8"	45½	80'-8"	60½	100'-8"	75½
1'-4"	1	21'-4"	16	41'-4"	31	61'-4"	46	81'-4"	61	101'-4"	76
2'-0"	1½	22'-0"	16½	42'-0"	31½	62'-0"	46½	82'-0"	61½	102'-0"	76½
2'-8"	2	22'-8"	17	42'-8"	32	62'-8"	47	82'-8"	62	102'-8"	77
3'-4"	2½	23'-4"	17½	43'-4"	32½	63'-4"	47½	83'-4"	62½	103'-4"	77½
4'-0"	3	24'-0"	18	44'-0"	33	64'-0"	48	84'-0"	63	104'-0"	78
4'-8"	3½	24'-8"	18½	44'-8"	33½	64'-8"	48½	84'-8"	63½	104'-8"	78½
5'-4"	4	25'-4"	19	45'-4"	34	65'-4"	49	85'-4"	64	105'-4"	79
6'-0"	4½	26'-0"	19½	46'-0"	34½	66'-0"	49½	86'-0"	64½	106'-0"	79½
6'-8"	5	26'-8"	20	46'-8"	35	66'-8"	50	86'-8"	65	106'-8"	80
7'-4"	5½	27'-4"	20½	47'-4"	35½	67'-4"	50½	87'-4"	65½	107'-4"	80½
8'-0"	6	28'-0"	21	48'-0"	36	68'-0"	51	88'-0"	66	108'-0"	81
8'-8"	6½	28'-8"	21½	48'-8"	36½	68'-8"	51½	88'-8"	66½	108'-8"	81½
9'-4"	7	29'-4"	22	49'-4"	37	69'-4"	52	89'-4"	67	109'-4"	82
10'-0"	7½	30'-0"	22½	50'-0"	37½	70'-0"	52½	90'-0"	67½	110'-0"	82½
10'-8"	8	30'-8"	23	50'-8"	38	70'-8"	53	90'-8"	68	110'-8"	83
11'-4"	8½	31'-4"	23½	51'-4"	38½	71'-4"	53½	91'-4"	68½	111'-4"	83½
12'-0"	9	32'-0"	24	52'-0"	39	72'-0"	54	92'-0"	69	112'-0"	84
12'-8"	9½	32'-8"	24½	52'-8"	39½	72'-8"	54½	92'-8"	69½	112'-8"	84½
13'-4"	10	33'-4"	25	53'-4"	40	73'-4"	55	93'-4"	70	113'-4"	85
14'-0"	10½	34'-0"	25½	54'-0"	40½	74'-0"	55½	94'-0"	70½	114'-0"	85½

FIGURE 11.9 Length of CMU walls by stretcher and height by coursing.
(continued on next page)

HEIGHT OF WALL	NO. OF UNITS	HEIGHT OF WALL	NO. OF UNITS	HEIGHT OF WALL	NO. OF UNITS	HEIGHT OF WALL	NO. OF UNITS	HEIGHT OF WALL	NO. OF UNITS	HEIGHT OF WALL	NO. OF UNITS
14'-8"	11	34'-8"	26	54'-8"	41	74'-8"	56	94'-8"	71	114'-8"	86
15'-4"	11½	35'-4"	26½	55'-4"	41½	75'-4"	56½	95'-4"	71½	115'-4"	86½
16'-0"	12	36'-0"	27	56'-0"	42	76'-0"	57	96'-0"	72	116'-0"	87
16'-8"	12½	36'-8"	27½	56'-8"	42½	76'-8"	57½	96'-8"	72½	116'-8"	87½
17'-4"	13	37'-4"	28	57'-4"	43	77'-4"	58	97'-4"	73	117'-4"	88
18'-0"	13½	38'-0"	28½	58'-0"	43½	78'-0"	58½	98'-0"	73½	118'-0"	88½
18'-8"	14	38'-8"	29	58'-8"	44	78'-8"	59	98'-8"	74	118'-8"	89
19'-4"	14½	39'-4"	29½	59'-4"	44½	79'-4"	59½	99'-4"	74½	119'-4"	89½
20'-0"	15	40'-0"	30	60'-0"	45	80'-0"	60	100'-0"	75	120'-0"	90

NOMINAL HEIGHT OF CONCRETE MASONRY WALLS BY COURSES
(Based on units 7⅝" high ⅜" thick mortar joints)

NO. OF UNITS	HEIGHT OF WALL	NO. OF UNITS	HEIGHT OF WALL	NO. OF UNITS	HEIGHT OF WALL	NO. OF UNITS	HEIGHT OF WALL
1	0'-8"	13	8'-8"	25	16'-8"	37	24'-8"
2	1'-4"	14	9'-4"	26	17'-4"	38	25'-4"
3	2'-0"	15	10'-0"	27	18'-0"	39	26'-0"
4	2'-8"	16	10'-8"	28	18'-8"	40	26'-8"
5	3'-4"	17	11'-4"	29	19'-4"	41	27'-4"
6	4'-0"	18	12'-0"	30	20'-0"	42	28'-0"
7	4'-8"	19	12'-8"	31	20'-8"	43	28'-8"
8	5'-4"	20	13'-4"	32	21'-4"	44	29'-4"
9	6'-0"	21	14'-0"	33	22'-0"	45	30'-0"
10	6'-8"	22	14'-8"	34	22'-8"	46	30'-8"
11	7'-4"	23	15'-4"	35	23'-4"	47	31'-4"
12	8'-0"	24	16'-0"	36	24'-0"	48	32'-0"

HOW TO USE THESE TABLES

The tables on this page are an aid to estimating and designing with standard concrete masonry units. The following are examples of how they can be used to advantage.

Example:
Estimate the number of units required for a wall 76' long and 12' high.
From table: 76' = 57 units
12' = 18 courses
57 × 18 = 1026 = No. masonry units required

Example:
Estimate the number of units required for a foundation 24' × 30' = 11 courses high.
2 (24 + 30) = 108' = distance for a foundation
From table: 108' = 81 units
81 × 11 = 891 = No. masonry units required.

This table can also be useful in the layout of a building on a modular basis to eliminate cutting of units. Example: If design calls for a wall 41' long it can be found from the table that making wall 41'-4", will eliminate cutting units and consequent waste. Example: If the distance between two openings has been tentatively established at 2'-9", consulting the table will show that 2'-8" dimension would eliminate cutting of units.

FIGURE 11.9 Length of CMU walls by stretcher and height by coursing (continued).

7. Is there accelerated corrosion from chlorides, or has galvanic action taken place because of carbon steel anchors in contact with dissimilar materials?

8. Has excessive moisture penetrated the wall system from any poorly maintained parapet or roof coping flashings?

9. Have any caulk joints been allowed to deteriorate? Are any actually missing?

10. Have the weep holes been caulked when maintenance caulking was performed? Have the lintels been caulked at the point where the brick is bearing on them?

11. Have the mortar joints deteriorated? Have they been tuck-pointed during routine maintenance inspections?

CLEANING BRICK WALLS

Cleaning brick walls is easier if certain precautions were taken as they were being built.

Learning how to build clean brick walls will reduce the time and expense required to clean them.

Efflorescence

That white, sometimes fluffy, substance that appears on recently completed masonry walls, known as efflorescence, is the leaching out of water-soluble salts contained in lime deposits in the mortar. These salts are called leachates, and while they do not affect the structural integrity of the wall, they may indicate that water is penetrating the wall at a somewhat higher elevation.

In new wall construction, efflorescence will occur as water passes through the masonry units and the mortar as it cures and brings dissolved salts with it, depositing these salts on the face of the brick or block. You will need to remove this whitish substance right after it first occurs and more than likely several more times until most of the moisture in the wall has evaporated.

Most efflorescence can be removed by dry brushing with a stiff bristled brush and then flushing with clean water. For persistent stains, a light sandblasting will work, but care has to be taken to avoid damaging the integrity of the mortar or the face of the masonry unit. There are also some specially formulated products on the market for the removal of efflorescence.

Brick Cleaning Systems

There are several brick cleaning systems, each with its own procedure. It is a good idea to become familiar with various methods. These include high-pressure water, sandblasting, wet cleaning, and specialty cleaning. Consult use guides and informational materials provided with the system being used.

12

THE BUILDING ENVELOPE

Roof, perimeter wall structure, and exterior wall treatments create the building envelope. Setting design criteria aside, the materials, shapes, textures, and colors for this phase of construction are limited only by one's flair and imagination—and pocketbook. Wood framing is the most common material employed in single family housing, but metal framing, the choice of commercial and low-rise apartment builders, is slowly gaining acceptance by residential builders. Engineered wood products (that's what plywood, "particle board," and Masonite are called nowadays), a category that includes such items as OSB (oriented strand board—the most commonly used exterior wall sheathing), "glulams" (laminated wood joists), I joists (lumber top and bottom chord with a particleboard web) and MDF (medium-density fiberboard) products, are now standard in the industry. Most roof and many floor structure systems are constructed with wood trusses, replacing the more labor-intensive roof rafter and floor joist systems.

FRAMING

Single-family residential construction follows two basic framing methods: standard or platform framing, sometimes referred to as western framing (See Figure 12.1 and Figure 12.2) and balloon

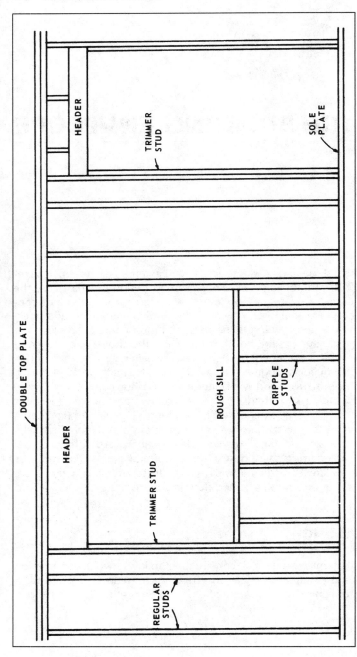

FIGURE 12.1 Framing detail.

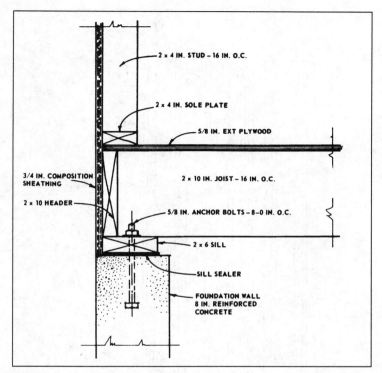

FIGURE 12.2 Framing detail.

framing (Figure 12.3). Platform framing is more common today; balloon framing was more prevalent prior to 1930.

In platform framing the first floor is built on top of the foundation, and the wall sections extend to the second-floor level (in the case of a two-story house). Floor joists and subflooring of either plywood or OSB are installed and nailed in place, and the exterior wall framing for the second floor uses the first-floor framing for bearing and its "platform".

Balloon framing consists of wall framing members that extend continuously to the roof rafter plate. The ends (connection points) of the second-floor joists are supported on a ribbon spiked to the vertical studs.

Framed wall openings in balloon framing differ from openings in western framing, reflecting the different ways in which structural stability is created.

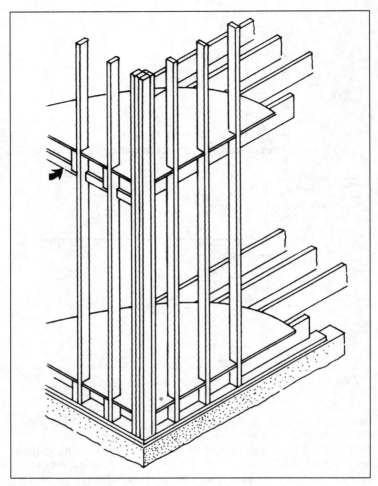

FIGURE 12.3   Balloon framing.

In either type of wall framing, openings for doors and windows are framed in basically the same manner; however, nomenclature of the framing members may vary from one geographic area to another. Some carpenters call cripples (short studs installed under a window opening) "cats," and the second supporting stud at a

fastfacts

If exterior wall sheathing is installed before all framing "dead loads" (weight) are created, the sheathing may begin to buckle if the joints are too tight.

door or window opening may be called a "jack" stud or simply a double stud.

Once the exterior walls are erected and the interior framing, subflooring and roof rafters are installed, the exterior wall sheathing can begin.

Figure 12.4 reveals a typical framing layout for a residential structure with members identified. Figure 12.5 is representative of a typical residential structure framed and sheathed with engineered wood products.

LIVE AND DEAD LOADS

A dead load can be defined simply as the load (weight) imposed by the structure itself; a live load is the added weight imposed by such elements as furniture, major appliances, and other significant pieces of equipment and by the weight of the occupants of the structure. There is another "load" factor that is also significant, wind load; the load, in this case *force,* imposed on the structure by the wind. Wind load considerations affect the design of exterior walls, which must be made "stiff" enough to withstand the forces imposed by the wind. Figure 12.6 is a cutaway of a residential structure showing what are typically considered dead and live loads

Exterior wall sheathing not only adds rigidity to the structure itself but provides a nailable surface to which wall finishes can be applied, whether wood or vinyl siding, brick on stone veneer, or synthetic stucco. The exterior wall also serves as a barrier to control moisture infiltration. Building paper, a black asphalt-impregnated material generally referred to as "15-pound felt," is frequently used where vinyl siding or brick or stone veneer is to be installed as an exterior wall treatment. However, a material called housewrap, more recognizable by the brand name Tyvek™, has

FRAME CONSTRUCTION

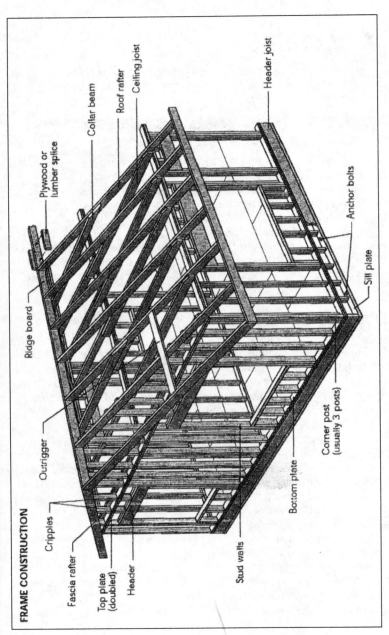

FIGURE 12.4 Typical framing layout for residential construction.

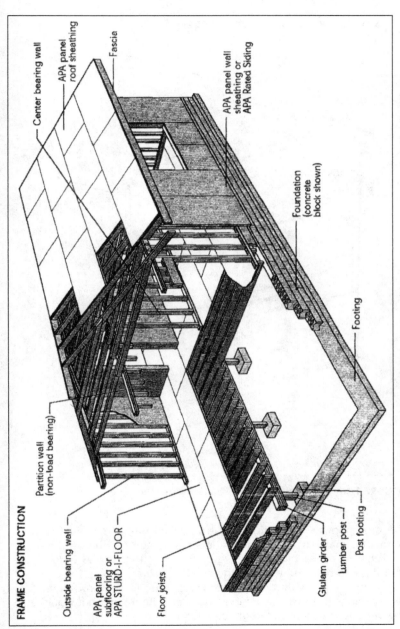

FRAME CONSTRUCTION

Center bearing wall

APA panel roof sheathing

Fascia

APA panel wall sheathing or APA Rated Siding

Foundation (concrete block shown)

Footing

Partition wall (non-load bearing)

Outside bearing wall

APA panel subflooring or APA STURD-I-FLOOR

Floor joists

Glulam girder

Lumber post

Post footing

FIGURE 12.5 Typical framed construction utilizing engineered wood products.

A Structure's Live and Dead Loads

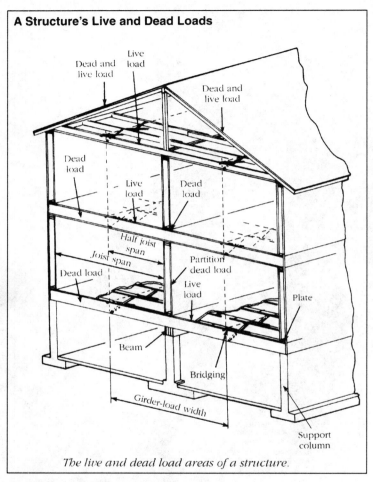

The live and dead load areas of a structure.

FIGURE 12.6 Typical dead and live loads in a structure.

become popular in providing control of moisture and air infiltration. This material significantly reduces air infiltration through the wall because it is made of a woven material that allows indoor humidity to dissipate through the exterior wall assembly.

Hand-in-hand with roofing materials and installation procedures are flashings, those metal or fabric materials without which it would be difficult to build a watertight structure.

WALL AND ROOF FRAMING MATERIALS

Metal and wood studs are the two most frequently used framing materials in residential and light commercial projects. When metal stud construction is employed, 18-, 20-, and 25-gauge (thickness) galvanized steel are the most common stud types used. The height of the wall and the loads imposed upon it (dead, live, and wind) will determine the correct size, length and gauge of the stud.

Framing Lumber

Both softwoods and hardwoods are used for framing. These terms may be somewhat misleading; they don't refer to the softness or hardness of the wood but rather to the type or species of tree from which the wood is taken. Softwoods come from trees such as evergreens that do not shed their leaves (cedar, pine, spruce, hemlock), while hardwoods come from trees that shed their leaves at the end of their growing season (oak, hickory, chestnut, maple, birch).

Softwoods are generally used for framing wood structures, but some types of pine are used for decorative purposes such as paneling, shelving, and cabinetry. Hardwoods are primarily used for flooring (oak, maple) or for furniture, cabinetry, and millwork.

The characteristics of wood vary from tree to tree as well as from one section to another in the same tree, so some means of classifying or grading lumber is necessary to attain some standard of uniformity. Two lumber associations control grading, specifications, and quality control. The Western Wood Products Association (www.wwpa.org) and the Southern Pine Council (www.sfpa.org) are the dominant lumber associations in this country.

The WWPA governs the following wood products:

- Douglas fir and western larch
- Hemlock fir (referred to as hem-fir), California red fir, grand fir, noble fir, Pacfic silver fir, white fir, and western hemlock
- Western cedars: western red, Alaska, and Port Orford cedar

Under the jurisdiction of WWPA are the following wood species: Englemann spruce, Sitka spruce, lodgepole pine, balsam fir, Jack pine, red pine, and western spruce.

The Western Wood Products Association establishes standards and regulates the grading of products within their jurisdiction. Western wood products can be divided into four broad categories:

- Rough lumber: lumber that has not been dressed but only sawn-edged and trimmed to the extent of showing saw marks on all four sides
- Dressed or surfaced lumber: lumber that has been run through a surfacing machine to achieve a smooth and uniform surface on one side (designated as S1S), two sides (S2S), one edge (S1E), two edges (S2E), all four sides (S4S), or any combination thereof
- Worked lumber: lumber that in addition to being dressed or surfaced has been matched, shiplapped, or tongue and grooved.
- Resawn lumber: lumber that is dressed before resawing and not afterward; thickness of resawn lumber is not uniform.

WWPA provides detailed information on notching and boring of western wood products (Figure 12.7) to ensure that the structural integrity of a framing member is not jeopardized when cuts are made to allow the passage of electrical, plumbing or HVAC work.

Southern Pine Council products are defined as those species of wood whose major range in the U.S. is south of the Mason-Dixon Line and east of the Great Plains. Known as "hard pines" (hard-needled), these trees make up the genus known as Pinus.

The four principal species of southern pine are longleaf, slash, shortleaf, and loblolly ; minor species are Virginia pine, pond pine, and pitch pine.

Knots, Splits, Shakes, and Checks

Natural wood products will not be of uniform consistency but will contain knots, splits, checks, and shakes, some of which are acceptable and some of which are not. The Western Wood Products Association has developed a series of standards known as grading rules for these imperfections.

Moisture Content

Wood is a hydroscopic material, one that easily soaks up and retains moisture. If you were to place a piece of wood under a microscope, you would see lots of hollow tubular cells that transport moisture and nutrients to the tree. These same hollow tubular cells are what makes lumber a good insulating material. When a tree is cut down and sawed into lumber, the moisture remains in the wood until it either evaporates or is "cooked" out by drying in a kiln.

Both the Western Wood Products Association and the Southern Pine Inspection Bureau, an organization separate from the Southern

Pine Council, have standards to designate the moisture content in lumber. The moisture content in lumber is the amount of water contained in the lumber expressed as a percentage of weight of wood from which the water has been removed. Dry lumber is defined as having a moisture content of 19% or less; lumber with a moisture content in excess of 19% is classified as "unseasoned" lumber.

When standard lumber is grade-stamped, the stamp includes the condition or "seasoning" of the lumber.

- MC 15: lumber surfaced with a moisture content of 15% or less
- KD-15: kiln-dried lumber, surfaced, with a moisture content of 15% or less (kiln-dried lumber is lumber that has been heat-seasoned in a chamber to produce a pre-determined moisture content)
- S-Dry: lumber surfaced with a moisture content of 19% or less
- KD: kiln-dried lumber with a moisture content of 19% or less
- S-GRN: unseasoned lumber with a moisture content in excess of 19%

Pressure-Treated Lumber

There are three basic categories of wood preservatives used in pressure-treated lumber:

- Waterborne preservatives: clean, odorless, chemically treated applications, EPA approved for both interior and exterior use without a sealer; most common waterborne preservative is chromated copper arsenate, known as CCA
- Creosote and creosote/coal tar mixtures: the black, smelly material commonly used on railroad ties, utility poles, and pilings
- Pentachlorophenol: referred to simply as Penta and used in industrial applications such as utility poles

fastfacts

Moisture-content designations apply only at the time of shipment. Lumber shipped unwrapped, in open containers, or stored in the open at the lumberyard will pick up moisture, and a 15% can easily turn into a 19%.

Notching & Boring Guidelines

Intended for use by residential builders, this WWPA TIP Sheet serves as a guide to code-allowed size and placement of cuts (notching and boring) in floor-joist and stud-wall framing members.

A number of problems can occur if cuts are made through framing members to make room for plumbing or electrical runs, ductwork, or other mechanical elements such as sound or security systems.

Whenever a hole or notch is cut into a member, the structural capacity of the piece is weakened and a portion of the load supported by the cut member must be transferred properly to other joists.

It is best to design and frame a project to accommodate mechanical systems from the outset, as notching and boring should be avoided whenever possible; however, unforeseen circumstances sometimes arise during construction.

If it is necessary to cut into a framing member, the following diagrams provide a guide for doing so in the least destructive manner.

Diagrams comply with the requirements of the three major model building codes: Uniform (UBC), Standard (SBC), and National (BOCA), and the CABO One- & Two-Family Dwelling Code.

FLOOR JOISTS

The following references are to actual, not nominal dimensions. (See Figure 1: *Placement of Cuts in Floor Joists* and Table 1: *Maximum Sizes for Cuts in Floor Joists.*)

Holes: Do not bore holes closer than 2" from joist edges, nor make them larger than 1/3 the depth of the joist.

Notches: Do not make notches in the middle third of the span where the bending forces are greatest. Notches should be no deeper than 1/6 the depth of the joist. Notches at the end of the joist should be no deeper than 1/4 the depth. Limit the length of notches to 1/3 of the joist's depth.

When a Notch Becomes a Rip

Codes do not address the maximum allowable length of a notch; however, the 1991 *National Design Specification (NDS)* does limit the maximum length of a notch to 1/3 the depth of a member.

It is important to recognize the point at which a notch becomes a rip, such as when floor joists at the entry of a home are ripped down to allow underlayment for a tile floor.

Ripping wide dimension lumber lowers the grade of the material, and is unacceptable under all building codes.

Table 1: Maximum Sizes for Cuts in Floor Joists

Joist Size	Max. Hole	Max Notch Depth	Max. End Notch
2x4	none	none	none
2x6	1-1/2"	7/8"	1-3/8"
2x8	2-3/8"	1-1/4"	1-7/8"
2x10	3"	1-1/2"	2-3/8"
2x12	3-3/4"	1-7/8"	2-7/8"

When a sloped surface is necessary, a non-structural member can be ripped to the desired slope and fastened to the structural member in a position above the top edge. Do not rip the structural member.

STUD WALLS

When structural wood members are used vertically to carry loads in compression, the same engineering procedure is used for both studs and columns. However, differences between studs and columns are recognized in the model building codes for conventional light-frame residential construction.

The term "column" describes an individual major structural member subjected to axial compression loads, such as columns in timber-frame or post-and-beam structures.

The term "stud" describes one of the members in a wall assembly or wall system carrying axial compression loads, such as 2x4 studs in stud wall that includes sheathing or wall board. The difference between columns and studs can be further described in terms of the potential consequences of failure.

Columns function as individual major structural members, consequently failure of a column is likely to result in partial collapse of a structure (or complete collapse in extreme cases due to the domino effect). However, studs function as members in a system. Due to the system effects (load sharing, partial composite action, redundancy, load distribution, etc.), studs are much less likely to fail and result in a total collapse than are columns.

Notching or boring into columns is not recommended and rarely acceptable; however, model codes establish guidelines for allowable notching and boring into studs used in a stud-wall system.

Figures 2 and 3 illustrate the maximum allowable notching and boring of 2x4 studs under all model codes except BOCA. BOCA allows a hole one third the width of the stud in all cases.

Bored holes shall not be located in the same cross section of a stud as a cut or notch.

For additional information on framing (and common framing errors), contact WWPA for reprints of the following articles written by Association field staff.

Field Guide to Common Framing Errors (JLC-2) reprinted from *Journal of Light Construction*: article focuses on most commonly-encountered job-site errors. 6 pgs.

Common Roof-Framing Errors (JLC-3) reprinted from *Journal of Light Construction:* focuses on problems and solutions with trusses, rafters, collar ties and structural ridges. 4 pgs.

Picture Perfect Framing (B-1) reprinted from *Builder Magazine:* discusses cantilevers, joist hangers, blocking, notching and boring, cathedral ceilings, and over cutting tapers. 4 pgs.

Article reprints are 75 cents each to cover postage and handling.

FIGURE 12.7 Proper notching and boring guidelines *(Courtesy of Western Wood Products Assoc., continued on next two pages).*

Fig. 1: Placement of Cuts in Floor Joists

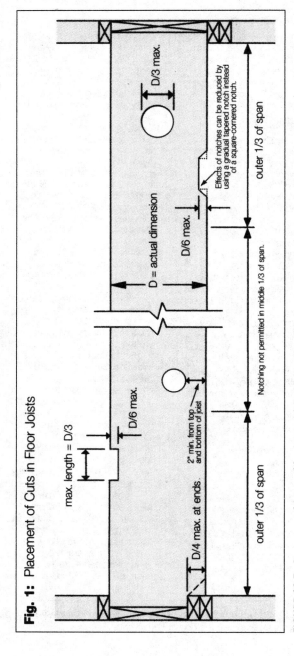

FIGURE 12.7 Proper notching and boring guidelines. *(continued)*

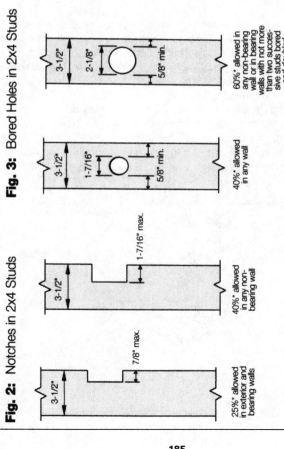

Fig. 2: Notches in 2x4 Studs

3-1/2" 7/8" max.

25%* allowed in exterior and bearing walls

3-1/2" 1-7/16" max.

40%* allowed in any non-bearing wall

Fig. 3: Bored Holes in 2x4 Studs

3-1/2" 1-7/16" 5/8" min.

40%* allowed in any wall

3-1/2" 2-1/8" 5/8" min.

60%* allowed in any non-bearing wall or in bearing walls with not more than two successive studs bored and doubled

Western Wood Products Association
522 SW Fifth Avenue Suite 400
Portland, OR 97204-2122
503/224-3930   Fax: 503/224-3934
e-mail: info@wwpa.org
web site: http://www.wwpa.org

WESTERN WOOD WORKS™

A-11/2017e/8-96, 2-97/15M

*Figures 2 and 3 illustrate 25%, 40% and 60% notches or holes in 2x4s (e.g. .25 x 3½" = .875 or 7/8"). These percentages apply to studs of any size.

FIGURE 12.7 Proper notching and boring guidelines. *(continued)*

185

Fasteners

The range of fasteners, nails, screws, nuts, and bolts is wide, with specialized products for just about every conceivable application. There are hammer-driven drive pins for fastening onto concrete and masonry products, toggle bolts for hollow wall fastening, powder-activated fasteners to be used by trained operators only, and screws of almost every imaginable shape, size, finish, and head design—not to mention nails.

ROOFS

Configurations

Gable roofs have a high point, the ridge, near the center of the roof; the roof slopes directly downward on both sides from this point. The gambrel roof, or barn roof, also has a central ridge. From this high point it pitches down equally on both sides to an intermediate point, from which it slopes again to the building's side walls; in effect it has two pitches.

A hip roof has a ridge; it does not extend from one end of the roof to the other, but short of the gable end, it pitches down on each end, creating a pitch on all four sides.

A shed roof has only one sloping plane, generally from front to back. A true mansard roof has a double slope on all four sides, but

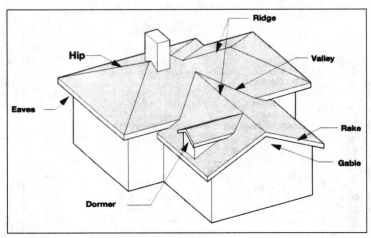

FIGURE 12.8 Common parts of a roof structure.

Defining Roof Slopes and other Types of Slopes

Percent Slope	Inch/Ft	Ratio	Degrees from Horizontal
1%	1/8	1 in 100	—
2%	1/4	1 in 50	—
3%	3/8	—	
4%	1/2	1 in 25	—
5%	5/8	1 in 20	3
6%	3/4	—	—
7%	7/8	—	—
8%	approx. 1	approx. 1 in 12	—
9%	1 1/8	—	
10%	1 1/4	1 in 10	6
11%	1 3/8	approx. 1 in 9	—
12%	1 1/2	—	
13%	1 5/8	—	—
14%	1 3/4	—	—
15%			8.5
16%	1 7/8	—	—
17%	2	approx. 2 in 12	—
18%	2 1/8	—	—
19%	2 1/4	—	—
20%	2 3/8	1 in 5	11.5
25%	3	3 in 12	14
30%	3.6	1 in 3.3	17
35%	4.2	approx. 4 in 12	19.25
40%	4.8	approx. 5 in 12	21.5
45%	5.4	1 in 22	24
50%	6	6 in 12	26.5
55%	6 5/8	1 in 1.8	28.5
60%	7 1/4	approx. 7 in 12	31
65%	7 3/4	1 in 1 1/2	33
70%	8 1/8	1 in 1.4	35
75%	9	1 in 1.3	36.75
100%	12	1 in 1	45

FIGURE 12.9   Defining roof slope as a percentage and inches per foot.

today's builders often refer to the vertically sloping front (generally shingled) of an otherwise flat roof as a "mansard" roof.

The hip roof diagram contains many of the roofing elements that apply equally to other roof configurations, such as ridge, valley, eave, and rake. Figure 12.8 illustrates and identifies the most common parts of a roof structure.

Many construction elements, such as shingles, rolled roofing, and built-up roofs, have limits based upon the degree of slope of the roof, referred to as pitch (expressed as a ratio). The chart in Figure 12.9 defines roof slope as a percentage, inches per foot, ratio, and degrees from horizontal.

A flat roof is defined as having a slope of ⅛ to 2 inches to the foot of horizontal (1% to 17%). Steep roofs are defined as having a slope of 2¼ to 12 inches per foot of horizontal (19% to 100%).

Trusses

There are a number of advantages in using trusses in roof construction; all the engineering is performed by the roof truss manufacturer to ensure that the design meets the needs of the building's dead and live loads. Truss-designed roof systems are not only cost-effective but high in quality; installation is less expensive than building trusses on the job site or in place.

There are standard roof-truss configurations and parallel chord trusses that can serve as a floor truss or joist or a flat roof truss (Figure 12.10). The range of truss configurations is limited only by the engineering imagination of their designers. Figure 12.11 contains diagrams of 23 different truss configurations; Figure 12.12 illustrates and identifies the various components of a truss.

Care must be taken during the unloading and erection process to avoid lateral bending or twisting. The Wood Truss Council of America has very specific instructions with respect to job handling, unloading, hoisting, and stacking of other materials. These unloading, stacking, lifting, bracing, and erection procedures should always be ovserved.

Roofing Materials

Asphalt, wood and slate shingles, and ceramic, metal, and composite tiles all find applicability on pitched roofs, and so does metal roofing, both flat panel and standing seam types.

Built-up roofing materials, rolled roofing, single-ply membrane, ballasted and mechanically fastened, fluid-applied, and foamed roofs are generally installed on flat roofs or nearly flat roofs of a pitch not recommended for shingles.

fastfacts

A "square" of shingles refers to the quantity required to cover a "square" of roof surface defined as an area 10' x 10' (100 SF).

TWO BASIC TYPES OF TRUSSES: The pitched or common truss is characterized by its triangular shape. It is most often used for roof construction. Some common trusses are named according to their web configuration, such as the King Post, Fan, Fink or Howe truss. The chord size and web configuration are determined by span, load, and spacing. All truss designs are optimized to provide the most economical application.

The parallel chord or flat truss gets its name from having parallel top and bottom chords. This type is often used for floor construction.

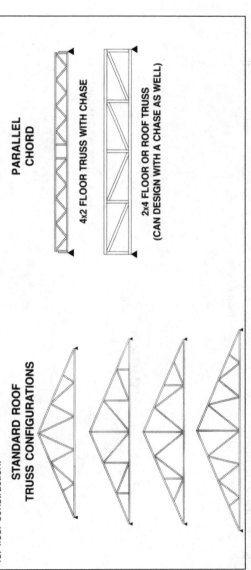

STANDARD ROOF TRUSS CONFIGURATIONS

PARALLEL CHORD

4x2 FLOOR TRUSS WITH CHASE

2x4 FLOOR OR ROOF TRUSS (CAN DESIGN WITH A CHASE AS WELL)

FIGURE 12.10 Standard and parallel roof and floor trusses *(Courtesy of Wood Truss Council of America, Madison, WI).*

TRUSS CONFIGURATIONS: The following examples represent some of the possible variations on the basic types of trusses. The only limit to the design is your imagination!

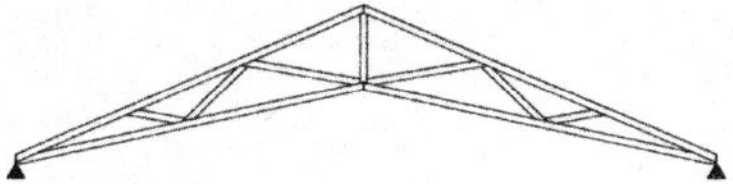

SCISSORS

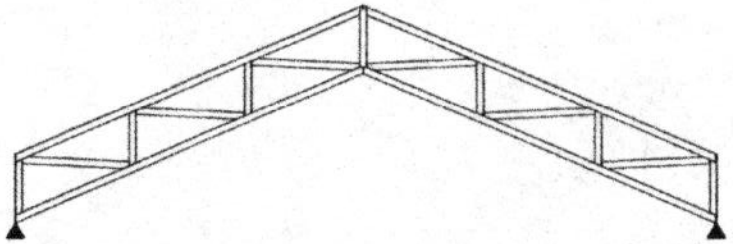

VAULTED PARALLEL CHORD

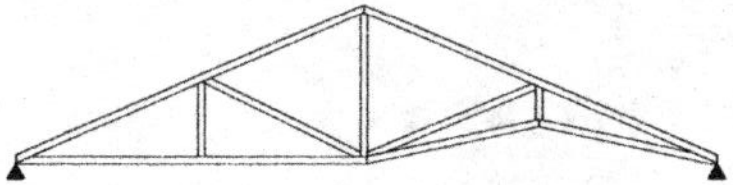

VAULT

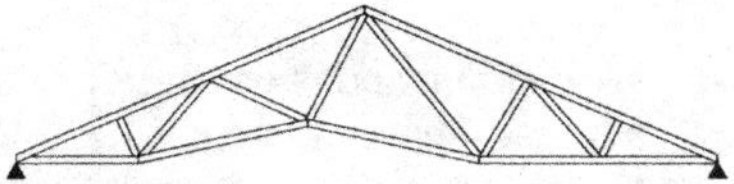

FLAT VAULT

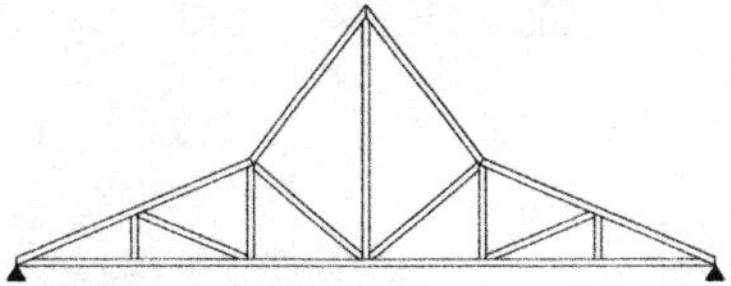

POLYNESIAN

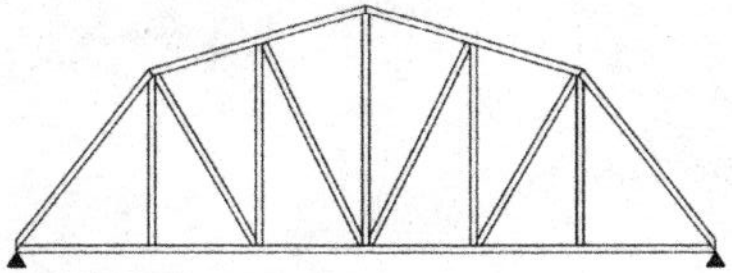

GAMBREL

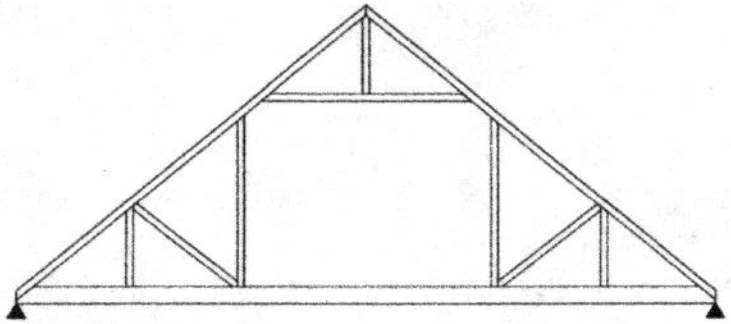

ROOM-IN-ATTIC

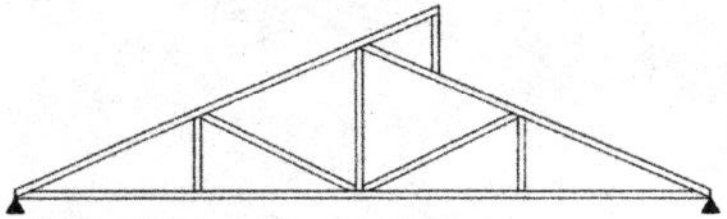

CLERESTORY

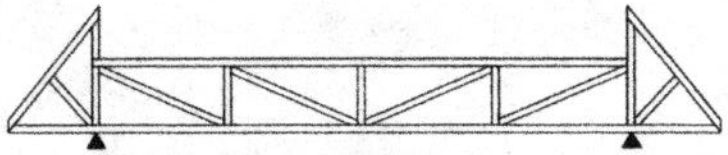

CANTILEVERED MANSARD W/PARAPETS

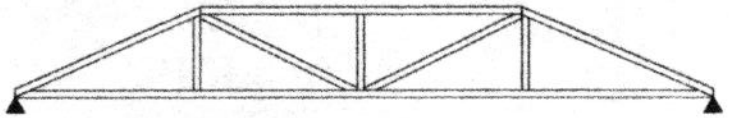

HIP

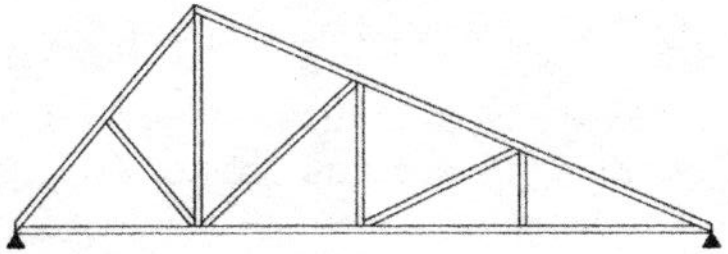

DUAL PITCH

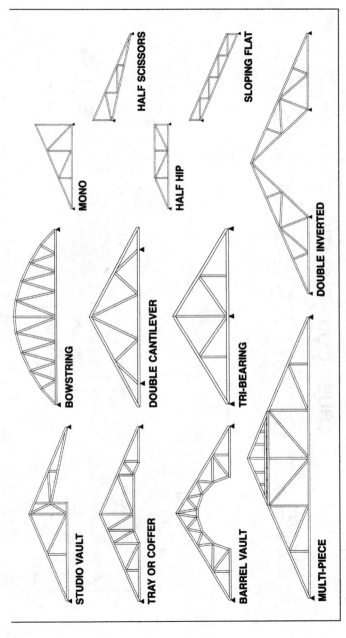

FIGURE 12.11 Twenty three roof truss configurations *(Courtesy of Wood Truss Council of America).*

TRUSS CONFIGURATIONS

TRUSS TERMS: The terms below are typically used to describe the various parts of a metal plate connected wood truss. The truss profile, span, heel height, overall height, overhang and web configuration depend on the specific design conditions and will vary by application.

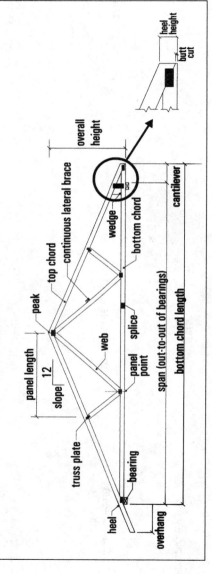

FIGURE 12.12 Nomenclature of roof truss components.

Asphalt Shingles

There are two basic types of asphalt shingles: fiberglass and organic. Fiberglass shingles are lighter in weight than organic, more resistant to heat and humidity and carry a Class A fire rating (the highest).

Organic shingles have a thick organic felt base saturated with a flexible asphalt material and then sealed on both side with a tougher asphalt material. The shingle is then covered with granules that resist abrasion from the elements. Shingles are divided into two further categories: standard and architectural or designer series. Architectural or designer shingles are generally of higher weight 425 pounds per square and offer many different color and shape options.

Standard shingles in the 225 to 245 pound per square range have 30-year limited warranties and offer a high quality product albeit in somewhat limited colors and sizes. Correct fastening of asphalt shingles is not a hit or miss situation; nails must be driven straight, first of all, and must not be under- or overdriven. The diagram in Figure 12.13 illustrates the proper nail set for asphalt shingles.

Wood Shingles and Shakes

Wood shingles and shakes (the latter are made from the same material but cut thicker and with more surface texture) are made from western red cedar, known for its long life exposed to the weather. Shingles and shakes will experience long life only if installed correctly; they must be allowed to "breathe" so that rainwater has a chance to evaporate. Installing them over 15-pound felt on spaced sheathing or 1- x 4-inch or 1- x 6-inch furring strips allows proper ventilation to occur particularly when the attic space over which the shingles have been installed have either louvers at the gable ends or ridge ventilation.

fastfacts

When installing wood shingles do not butt them tightly together; builders often keep them at least one nail width apart and the Cedar and Shake Institute guidelines are ¼ inch apart.

Example of fastener set for shingle nails.

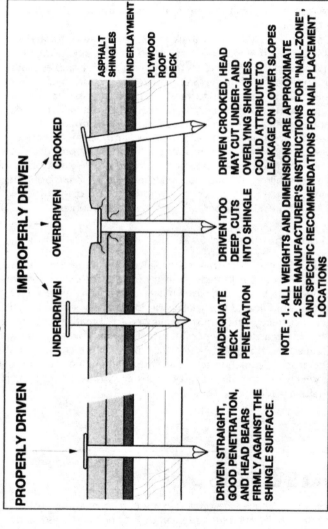

FIGURE 12.13   Properly and improperly driven fasteners.

Wood shingles and shakes are available in many different configurations. The following charts (Figures 12.14, 12.15, and 12.16) illustrate the type, length, and thickness, and approximate coverage for the most commonly produced shingles and shakes.

Slate Shingles

Slate, when installed and maintained properly, will last several lifetimes. Not only are the materials expensive and qualified installers hard to find, but the added weight of the shingles will require a stronger than average roof framing system.

Slate is quarried and cut into large chunks the size (length and width) of the finished shingle size. Hand splitting of this large chunk will create shingles of varying thickness ready to be slightly trimmed and drilled for fastening.

Nowadays, because of the high cost of a new slate roof, most slate roofing work is relegated to repair and partial replacement. Figure 12.17 lists standard slate shingle sizes, thickness, and weight per square.

Built-up Roofs

The typical (some say old-fashioned) method of roofing application, a series of fiberglass-reinforced/asphalt-impregnated sheets (plys) are unrolled over the roof deck and each mopped with a layer of hot asphaltic material. Two, three, or four plies of roofing "felts" are typically used over some form of rigid insulation, providing a tight, energy-efficient method of sealing.

Rolled Roofing

Just as the name implies, this roofing material comes in rolls, about 3 feet wide by 25 feet in length, and can be applied over a

fastfacts

When installing new or replacement slate shingles, proper nailing is essential. Don't drive the nail home too far; the nail head should just touch the shingle.

WOOD SHINGLES

NAME	LENGTH, THICKNESS, AND REFERENCE NOMENCLATURE	DESCRIPTION
NO.1** **TAPER-SAWN** 	16" x .40" FIVEX (406mm x 10mm) 18" x .45" PERFECTIONS (457mm x 11mm) 24" x .50" ROYALS (610mm x 13mm)	TOP GRADE (NO. 1**) WOOD SHINGLES FOR USE AS ROOFING. THE WOOD BLANKS ARE RUN DIAGONALLY THROUGH A BANDSAW TO PRODUCE SHINGLES THAT ARE SAWN BOTH SIDES.
NO.1** **FANCY SAWN BUTT** 	16" x .40" (406mm x 10mm) 18" x .45" (457mm x 11mm) 24" x .50" (610mm x 13mm)	TOP GRADE (NO. 1**) WOOD SHINGLE FOR USE AS ROOFING. TAPER-SAWN BOTH SIDES, WITH BUTT END CUT TO SPECIFIC SHAPE. A VARIETY OF FANCY BUTTS ARE AVAILABLE.

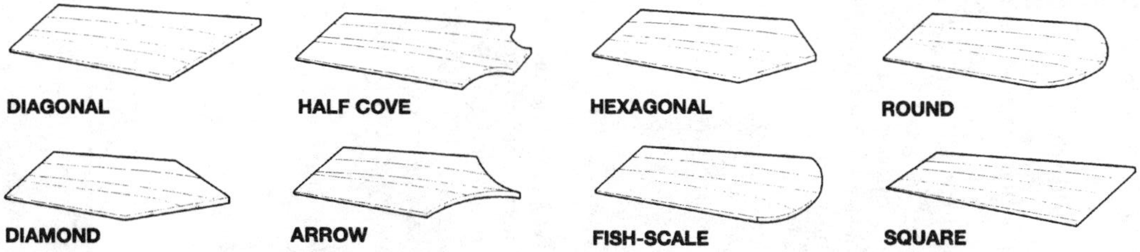

OCTAGON

DIAGONAL

HALF COVE

ROUND

DIAMOND

ARROW

FISH-SCALE

HEXAGONAL

SQUARE

** LOWER GRADES (e.g., NO. 2 AND NO. 3) ARE AVAILABLE AND ARE USED FOR STARTER OR UNDERCOURSING.

FIGURE 12.14   Wood shingle shapes and sizes.

Shingle coverage and exposure table.

SHINGLE COVERAGE

SHINGLES	LENGTH OF NO. 1 SHINGLES IN INCHES (mm)	APPROXIMATE COVERAGE IN SQ. FT. (m²) OF ONE SQUARE (4 BUNDLES) OF SHINGLES BASED ON FOLLOWING WEATHER EXPOSURES IN INCHES (mm)								
		3 1/2" (89)	4" (102)	4 1/2" (115)	5" (127)	5 1/2" (140)	6" (152)	6 1/2" (165)	7" (178)	7 1/2" (191)
 NO.1	16" (406)	70 (6.50)	80 (7.43)	90 (8.36)	100* (9.29)	—	—	—	—	—
	18" (457)	—	72.5 (6.74)	81.5 (7.57)	90.5 (8.40)	100* (9.29)	—	—	—	—
	24" (610)	—	—	—	—	73.5 (6.83)	80 (7.43)	86.5 (8.04)	93 (8.64)	100* (9.29)

* MAXIMUM EXPOSURE RECOMMENDED FOR ROOFS

FIGURE 12.15   Wood shingle coverage.

SHAKE COVERAGE

NO. 1 GRADE SHAKE TYPE, LENGTH, AND THICKNESS IN INCHES (mm)		APPROXIMATE COVERAGE IN SQ. FT. (m²) FOR 5 BUNDLES WHEN SHAKES ARE APPLIED WITH AN AVERAGE 1/2" (13mm) SPACING AT THE FOLLOWING WEATHER EXPOSURES, IN INCHES (mm)				
		5" (127)	5 1/2" (140)	7 1/2" (191)	8 1/2" (216)	10" (254)
SHAKES  NO.1 HANDSPLIT & RESAWN	HEAVIES 24" X 3/4" (610 x 19)	—	—	75(b) (6.96)	85 (7.90)	100(c) (9.29)
	MEDIUMS 24" X 1/2" (610 x13)	—	—	75(b) (6.96)	85 (7.90)	100(c) (9.29)
	HEAVIES 18" X 3/4" (457 x 19)	—	55(b) (5.11)	75(c) (6.96)	—	—
	MEDIUMS 18" X 1/2" (457 x 13)	—	55(b) (5.11)	75(c) (6.96)	—	—
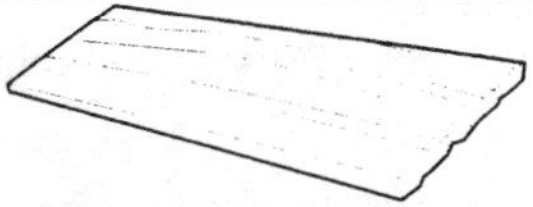	24" X 5/8" (610 x 16)	—	—	75(b) (6.96)	85 (7.90)	100(c) (9.29)
NO.1 TAPER-SAWN	18" X 5/8" (457 x 16)	—	55(b) (5.11)	75(c) (6.96)	—	—

FIGURE 12.16 Wood shaketypes, sizes, and coverage. *(continued on next page)*

Shake type	Size					
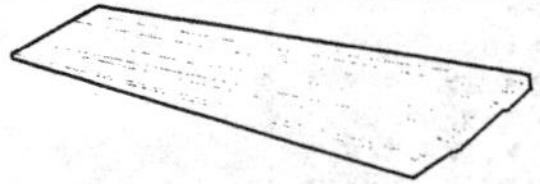 NO.1 TAPER-SPLIT	24" X 1/2" (610 x 13)	—	—	75(b) (6.96)	85 (7.90)	100(c) (9.29)
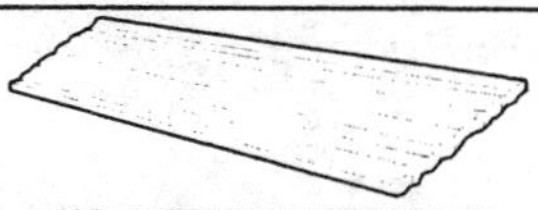 NO.1 STRAIGHT-SPLIT	18" X 3/8" (457 x 10)	—	65(b) (6.04)	90(c) (8.36)	—	—
HANDSPLIT STARTER	24" X 3/8" (610 x 10)	50 (4.65)	—	75(b) (6.96)	—	—
15" STARTER COURSE 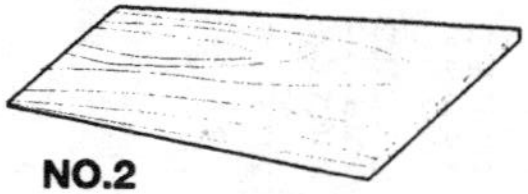	15" X 3/8" (381 x 10)	USE WITH SHAKES APPLIED NOT OVER 7 1/2" (191mm) WEATHER EXPOSURE				
NO.2	24" X 3/8" (610 x 10)	USE WITH SHAKES APPLIED NOT OVER 10" (254mm) WEATHER EXPOSURE				

(a) 5 BUNDLES MAY COVER 100 SQ. FT. (9.29m) WHEN USED AS STARTER COURSE AT 10" (254mm) WEATHER EXPOSURE; 7 BUNDLES MAY COVER 100 SQ. FT. (9.29m) WHEN USED AS STARTER COURSE AT 7 1/2" (191mm) WEATHER EXPOSURE.

(b) MAXIMUM RECOMMENDED WEATHER EXPOSURE FOR TRIPLE COVERAGE ROOF CONSTRUCTION.

(c) MAXIMUM RECOMMENDED WEATHER EXPOSURE FOR DOUBLE COVERAGE ROOF CONSTRUCTION.

(d) MAXIMUM RECOMMENDED WEATHER EXPOSURE.

NOTE - ALL DIMENSIONS ARE APPROXIMATE

Shake coverage and exposure table.

FIGURE 12.16 Wood shaketypes, sizes, and coverage. *(continued)*

SCHEDULE FOR STANDARD 3/16"(5mm) THICK SLATE

SIZE OF SLATE (L x W) (IN.)	SIZE OF SLATE (L x W) (MM)	SLATES PER SQUARE	EXPOSURE WITH 3"(76mm) Lap (IN.)	EXPOSURE WITH 3"(76mm) Lap (MM)	SIZE OF SLATE (L x W) (IN.)	SIZE OF SLATE (L x W) (MM)	SLATES PER SQUARE	EXPOSURE WITH 3"(76mm) Lap (IN.)	EXPOSURE WITH 3"(76mm) Lap (MM)
26 x 14	660 x 356	89	11 1/2	292	16 x 14	406 x 356	160	6 1/2	165
					16 x 12	406 x 305	184	6 1/2	165
24 x 16	610 x 406	86	10 1/2	267	16 x 11	406 x 279	201	6 1/2	165
24 x 14	610 x 356	98	10 1/2	267	16 x 10	406 x 254	222	6 1/2	165
24 x 13	610 x 330	106	10 1/2	267	16 x 9	406 x 229	246	6 1/2	165
24 x 12	610 x 305	114	10 1/2	267	16 x 8	406 x 203	277	6 1/2	165
24 x 11	610 x 279	138	10 1/2	267					
					14 x 12	356 x 305	218	5 1/2	140
22 x 14	559 x 356	108	9 1/2	241	14 x 11	356 x 279	238	5 1/2	140
22 x 13	559 x 330	117	9 1/2	241	14 x 10	356 x 254	261	5 1/2	140
22 x 12	559 x 305	126	9 1/2	241	14 x 9	356 x 229	291	5 1/2	140
22 x 11	559 x 279	138	9 1/2	241	14 x 8	356 x 203	327	5 1/2	140
22 x 10	559 x 254	152	9 1/2	241	14 x 7	356 x 178	374	5 1/2	140
20 x 14	508 x 356	121	8 1/2	216	12 x 10	305 x 254	320	4 1/2	114
20 x 13	508 x 330	132	8 1/2	216	12 x 9	305 x 229	355	4 1/2	114
20 x 12	508 x 305	141	8 1/2	216	12 x 8	305 x 203	400	4 1/2	114
20 x 11	508 x 279	154	8 1/2	216	12 x 7	305 x 178	457	4 1/2	114
20 x 10	508 x 254	170	8 1/2	216	12 x 6	305 x 152	533	4 1/2	114
20 x 9	508 x 229	189	8 1/2	216					
					11 x 8	279 x 203	450	4	102
18 x 14	457 x 356	137	7 1/2	191	11 x 7	279 x 178	515	4	102
18 x 13	457 x 330	148	7 1/2	191					
18 x 12	457 x 305	160	7 1/2	191	10 x 8	254 x 203	515	3 1/2	89
18 x 11	457 x 279	175	7 1/2	191	10 x 7	254 x 178	588	3 1/2	89
18 x 10	457 x 254	192	7 1/2	191	10 x 6	254 x 152	686	3 1/2	89
18 x 9	457 x 229	213	7 1/2	191					

Schedule for standard slate.

Slate Weight

A square of slate on the roof (i.e., enough slate to cover 100 square feet [9m²] of roof surface, set at the standard 3 inch [76 mm] headlap) will vary in weight from around 650 to 8,000 pounds [295 to 3,629kg], depending on the thickness of each slate (from ³⁄₁₆ inch to 2 inches [4mm to 51mm]).

Slates of commercial standard thickness (approximately ³⁄₁₆ inch [4mm]) weigh approximately 850 pounds per square (41kg/m²).

Slate products vary significantly in weight because of the numerous different sizes (lengths and widths), thickness, and types available for use from the different quarries. Table 2 shows the approximate weights (per square [kg/m²]) of roofing slates of different thicknesses. The actual weight of slate products may be from 10 percent above to 15 percent below the weights shown in the table.

Slate Thickness	Sloping Roof With 3'' (76mm) Lap (approx. Pounds Per Square [kg/m²])
³⁄₁₆'' to ¼'' (4mm to 6mm)	700 to 1,000 lbs/sq (3,417 to 4,882 kg/m²)
³⁄₈'' (9mm)	1,500 lbs/sq (7,323 kg/m²)
½'' (13mm)	2,000 lbs/sq (9,764 kg/m²)
¾'' (19mm)	3,000 lbs/sq (14,646 kg/m²)
1'' (25mm)	4,000 lbs/sq (19,528 kg/m²)
1¼'' (32mm)	5,000 lbs/sq (24,410 kg/m²)
1½'' (38mm)	6,000 lbs/sq (29,292 kg/m²)
1¾'' (44mm)	7,000 lbs/sq (34,174 kg/m²)
2'' (51mm)	8,000 lbs/sq (39,056 kg/m²)

FIGURE 12.17 Slate shingle size, weight, and exposures.

plywood or particleboard substrate deck with a roof slope as low as 1 to 6 inches per foot. Rolled roofing can also be applied over a concrete substrate if the concrete is fully cured and has been primed with an appropriate primer . Most manufacturer's instructions suggest priming the roof deck with an asphaltic primer after the old roof materials have been removed and the deck swept. The rolled roofing, with an adhesive backing, is then unrolled on the deck. Subsequent sheets are overlapped and joints sealed with roof mastic; end and side laps are also sealed with roof mastic, while some types of rolled roofing are heat-sealed at the joints.

Single Ply Membrane

Just as the name implies, a single-ply membrane made of PVC or EPDM (a rubber compound referred to chemically as ethylene propylene diene terpolymer) is rolled out over the roof deck (generally a flat deck), seams are lapped, and solvent welded in the case of PVC, and "glued" with a special rubber-cement material in the case of EDPM. Some materials are available in 40-foot-wide rolls so that seams are reinforced and potential sources of leaks are minimized. Single-ply roofs are installed in one of two fashions: mechanically fastened to the substrate with specifically designed screws or ballasted, in which a layer of round, preferably river stone devoid of sharp edges is placed over the entire roof surface. Of course, the structural system supporting the roof must be substantial enough to bear the added weight of the stone when the ballasted method is selected.

Metal Roofs

Tin roofs were used by farmers for years. They were inexpensive and lasted a long time. Metal and coating technology has developed quite a bit from those shiny tin roof days, and galvanized metal or aluminum panels with baked-on fluorocarbon paints such as PPG's Kynar™ promise a 20-year life against fading or chalking. Metal roofing materials are available in flat-panel or raised-standing-seam configuration.

Roof Tiles

Replicating Spanish-style roofs popular in the Southwest, these curved roof sections are still made from clay, but authentic-looking Spanish tiles are also made from metal coated with a fired-on material to simulate clay. Various configurations of clay, cement, or coated base metal tiles are popular, primarily in the southern and southwestern United States. Some of these cement-based products are also manufactured to look like wood shingles or wood shakes.

Fluid-Applied Membranes

There are a number of fluid-applied roof membrane compounds on the market today. Basically, they are bitumastic base materials that are flooded onto the roof substrate in liquid form and squeegeed over the entire surface to form a uniform liquid membrane, which when cured is watertight. The term "cured" refers to the chemical process whereby the flooded liquid changes to a flexible waterproof membrane.

This type of roof is generally installed over a concrete roof deck, thereby assuring that significant movement in the substrate that could cause tears in the membrane will be avoided. However, there is enough elasticity in the membrane to bridge small shrinkage cracks that might appear in the concrete deck.

Foamed Roofing

Back in the 1970s and 1980s when energy conservation became a hot topic, several manufacturers developed a sprayed-on foam that hardened as it cured. When sprayed on a roof deck to a depth of 4 to 8 inches, very high "R" (insulating) values could be achieved. This system fell of out favor as more sophisticated types of insulation were developed, and the single-ply membrane roof took on more popularity in commercial and industrial applications due to ease of installation, little maintenance, and cost effectiveness.

Flashings

Flashings are a necessary and critical component of a roof, preventing the intrusion of water into the structure it protects. Copper, aluminum, galvanized steel, and lead are the most popular forms of flashing materials used in roof construction.

Typical roof flashings are:

- Gravel stop: metal flashings attached to the edge of a roof to protect and secure the edge of the roof membrane.
- Copings: similar in nature to gravel stops except they are placed on the top of roof parapet walls.

- Base flashings: generally flexible materials that provide watertight integrity between a horizontal roof membrane and some vertical surface; they can be made of metal, requiring a reglet (a slit created usually in a masonry wall to receive and secure the vertical flashing), or counterflashing on the vertical surface to ensure watertight integrity.

- Counterflashing: flashings that act as a shield to cover the base flashing below.

- Pipe and conduit: whenever a pipe or conduit for electricity, HVAC, or plumbing penetrates the roof, flashings are required to seal off the penetration (also see pitch pocket below).

- Roof drain flashings: if internal rainwater drains are installed on a roof (primarily a flat roof), generally at a low point on the roof flashings especially designed for this purpose are installed.

- Roof vent flashings: roof vents and exhaust fan ducts installed through the roof sealed with special types of flashing designed for that purpose; they are usually bought as an accessory to the actual roof vent.

- Pitch pockets: the "pocket," a square metal box, is formed of aluminum or copper and fastened to the roof structure to enclose one or more pipes or conduits that penetrate the roof; after these pipes are run through the roof and into the pitch pocket, the "pocket" is filled with a black viscous tar that "cold"-flows to seal off any spaces around the pipe penetrations; pitch pockets require periodic inspection to insure that levels of "pitch" are in good condition.

- Ridge flashings: where the valley of a roof meets the eaves, generally on shingled roofs, flashings are installed.

Gutters and Downspouts

Although familiar to any homeowner, what may not be so well known is the terminology of the various parts and attachments that make up a typical gutter and downspout installation.

Figure 12.18 depicts a typical metal roof gutter and downspout installation with all parts identified. Figure 12.19 contains various gutter and downspout construction details.

Roof Troubleshooting Tips

1. Find the problem (such as ceiling spots or water stains)
   Check attic, chimney, roof vents for cracks in flashings or

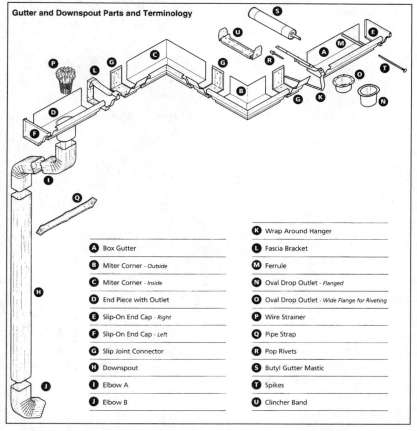

Gutter and Downspout Parts and Terminology

A Box Gutter	**K** Wrap Around Hanger
B Miter Corner - *Outside*	**L** Fascia Bracket
C Miter Corner - *Inside*	**M** Ferrule
D End Piece with Outlet	**N** Oval Drop Outlet - *Flanged*
E Slip-On End Cap - *Right*	**O** Oval Drop Outlet - *Wide Flange for Riveting*
F Slip-On End Cap - *Left*	**P** Wire Strainer
G Slip Joint Connector	**Q** Pipe Strap
H Downspout	**R** Pop Rivets
I Elbow A	**S** Butyl Gutter Mastic
J Elbow B	**T** Spikes
	U Clincher Band

FIGURE 12.18 Typical metal roof gutter with components identified.

cracked or missing caulk joints. Check roof rafters for evidence of leaks that could have traveled away from their original source. Check roof for damaged or missing shingles or pieces that could have cracked or split and fallen away.

2. Inspect flashings: are they improperly installed and separating from chimney or valleys? Check for cracks in metal or nails that may have popped as the wood underneath shrunk. Check all caulked joints.

3. Check roof surface for any missing shingles that may have been improperly fastened or damaged by exposure to heavy

Gutter/Downspout Accessories Typical Construction Details

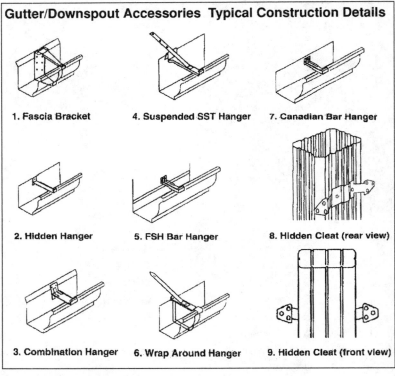

1. Fascia Bracket
4. Suspended SST Hanger
7. Canadian Bar Hanger

2. Hidden Hanger
5. FSH Bar Hanger
8. Hidden Cleat (rear view)

3. Combination Hanger
6. Wrap Around Hanger
9. Hidden Cleat (front view)

FIGURE 12.19 Various gutter and downspout details *(continued on next page).*

winds. Check for buckling and curling caused by improperly installed felts or by movement of the substrate (plywood deck). Look at the surface of the shingles. Blistering is caused by moisture under the shingles or from application of too much asphaltic cement. Check for missing granules; on newly

fastfacts

When you spot roof damage, check with your insurance agent; your policy may cover some or all of the costs.

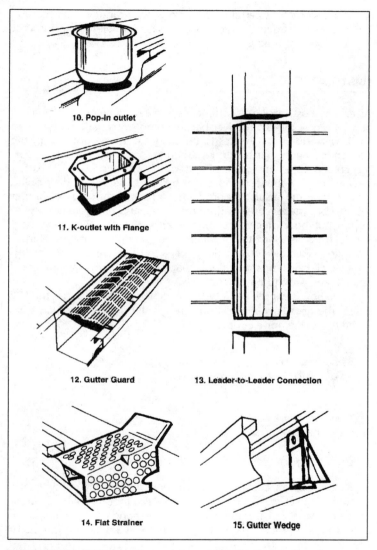

10. Pop-in outlet

11. K-outlet with Flange

12. Gutter Guard

13. Leader-to-Leader Connection

14. Flat Strainer

15. Gutter Wedge

FIGURE 12.19 Various gutter and downspout details. *(continued)*

installed roofs, a certain amount of loose granules will accumulate. But on aged shingle roofs, missing granules are a sign that replacement time is near.

INSULATION

As energy costs increased dramatically over the years and the concept of renewable alternatives became a hot topic, attention turned to conservation. As it relates to building and home ownership, the cost effectiveness of insulating a structure to retain internal heat during cold weather and ward off exterior heat during warm weather became readily apparent. Advances by manufacturers in developing a wide range of insulating products and the technology of proper installation make conservation measures affordable and easy to apply. But how much insulation is enough and where and how should it be installed? Batt insulation, the predominant type of insulation in residential construction, needs to be installed in attics, ceilings, wall cavities, some floor spaces (particularly over a crawl space), and around basement walls.

Insulation is expressed in terms of "R" values. The higher the "R" value, the greater the insulating quality. All building materials have some "R" value, for example:

Material	R Value
8" concrete block	1.11
⅝" drywall	0.56
½" plywood	0.02
⅛" floor tile	0.05
1" wood core door	1.96
single pane of glass	0.94

Choosing Insulation

Batt insulation can be obtained with no backing or with aluminum-foil-backed paper that functions as a moisture barrier. Available in a variety of thicknesses with corresponding "R" values, the size of framing members in the structure will limit the thickness of batt insulation to be installed.

- 2 x 4 framing will accommodate R-11, R-13, R-15 batts.
- 2 x 6 framing will accommodate R-19 and R-21 batts.
- 2 x 10 framing members will accommodate R-30 and R-38 batts.
- Flat surfaces of an attic will accommodate R-49 or R-60 batts.

What are considered adequate and superior insulating values for roofs and walls? This of course varies from geographic area to geographic area based upon the extremes of climate, but the following guidelines are worth considering:

In New England
Attic	R 49 is good	R-60 is best
Wall cavities	R-19 is good	R-21 is best
Floor	R-30 is good	R-38 is best
Basement walls	R-13 is good	R-19 is best
Crawlspace wall	R-19 is good	R-30 is best

In Mid-Atlantic
Attic	R-38 is good	R-60 is best
Wall cavities	R-19 is good	R-21 is best
Floor	R-19 is good	R-38 is best
Basement walls	R-13 is good	R-19 is best
Crawlspace wall	R-19 is good	R-30 is best

In Southeast
Attic	R-30 is good	R-38 is best
Wall cavities	R-13 is good	R-15 is best
Floor	R-11 is good	R-19 is best
Basement walls	R-11 is good	R-15 is best
Crawlspace wall	R-11 is good	R-19 is best

Midwest and Upper Midwest
Attic	R-38 is good	R-60 is best
Wall cavities	R-19 is good	R-21 is best
Floor	R-19 is good	R-38 is best
Basement walls	R-13 is good	R-19 is best
Crawl pace wall	R-19 is good	R-30 is best

Southwest
Attic	R-30 is good	R-49 is best
Wall cavities	R-13 is good	R-19 is best
Floor	R-13 is good	R-30 is best
Basement walls	R-11 is good	R-15 is best
Crawlspace wall	R-11 is good	R-19 is best

Mid- to Northern California
Attic	R-30 is good	R-38 is best
Wall cavities	R-13 is good	R-15 is best
Floor	R-11 is good	R-19 is best
Basement walls	R-11 is good	R-15 is best
Crawlspace wall	R-11 is good	R-19 is best

fastfacts

Two installation tips:

1. *Don't compress batts in the opening; it decreases their efficiency.*

2. *When using foil or paper-faced batts, make sure the tabs overlap so that the integrity of the vapor barrier is assured.*

Batt Size and Compatibility

"R" value	Thickness	Use in
R-15	3½ inches	2 x 4 framing
R-19	5½ inches	2 x 6 framing
R-21	6¼ inches	2 x 6 framing
R-30	9½ inches	Roof framing
R-38	12 inches	Roof framing
R-38C	10¼ inches	2 x 12 cathedral ceilings

Roof assemblies must take many other factors into consideration:

- Climate: exterior and interior temperatures and humidity
- Building and roof life expectancy
- Type of roof deck
- Architectural design-slope and drainage considerations
- Code requirements
- Thermal/energy conservation requirements
- Fire, wind, and impact resistance criteria

Albedo

Albedo is the measurement of the energy efficiency of roof membranes. Resistance to heat flow has been quantified by the use of "R" values. But when it comes to energy lost or gained through a roof assembly, another measurement is often used, referred to as "albedo," solar reflectance. Albedo measures how much of the solar energy striking a roof surface is reflected off the roof.

Energy-efficient roofing systems exhibit three qualities:

- Good reflectivity—*albedo*
- Sufficient *insulation*—to resist the flow of heat into the structure in summer and resist the flow of heat out of the structure in winter
- Good *emissivity*—the ability of the roof surface to radiate the absorbed energy away from the structure rather than retain it

The following chart lists albedo and emissivity factors for selected surfaces (the higher the Albedo number, the better the reflectivity).

Material	Albedo	Emissivity
Concrete	0.3	0.94
Red brick	0.3	0.90
Tar paper	0.05	0.93
White plaster	0.93	0.91
Bright galvanized steel	0.35	0.13
Bright aluminum foil	0.85	0.04
White pigment	0.85	0.96
White single-ply roofing	0.78	0.90

SEALANTS

The category of sealants spans a wide variety of construction materials from those used to prevent water from infiltrating into below-grade structures and basements to masonry sealers to joint filling compounds.

These materials generally fall into one of three categories:

1. Dynamic joint fillers: materials used to prevent water infiltration into joints that move from expansion and contraction or from applied loadings. Expansion joint materials placed in concrete sidewalks are a good example of this type of filler.

fastfacts

Whenever you experience a roof leak, check the flashings first!

2. Static joint fillers: materials used to fill joints that exhibit little or no movement, such as masonry control joints or cracks in masonry joints. However, no joint in a building is completely static because all materials exhibit some movement from temperature and loading changes.

3. Butt joints: fillers that bridge opposing faces that contract and expand, placing a sealant in compression and tension and also possibly exhibiting shear from extreme loading forces or seismic events.

The key to proper application of many sealants begins with proper surface preparation. Most manufacturers go to great lengths to provide detailed surface preparation information along with their application procedures. Too often, applicators ignore the preparation instructions and, when failure of the joint occurs, are quick to blame the product. Before applying any joint sealer, the first step is to thoroughly read and understand the preparation requirements. The second step is to follow them!

The following guidelines are general in nature and should be supplemented by the specific manufacturer's requirements generally contained on the product itself:

- Concrete and masonry: concrete can have the most variable surface conditions of any product because of placement and finishing techniques, curing procedures, additives to the concrete itself, and form-release agent residues. Concrete and some masonry surfaces can exhibit weak surface layers because of laitance present in the concrete or mortar. Surfaces contaminated by laitance, hardeners, curing compounds, and form-release materials can be lightly sandblasted or wire-brushed away. Newly placed concrete or masonry must be allowed to thoroughly cure before applying any sealants. If these cured surfaces are exposed to rain or surface water, they should be allowed to dry for 24 hours in good drying weather before sealer or primer application. Because most sealant manufacturers do not recommend applying their product in temperatures below 40 degrees F, frost on the surface presents a problem. Under light frost conditions an application of isopropyl alcohol or methyl ethyl ketone will cause surface moisture to evaporate and a sealant can be quickly applied—if ambient temperatures are within an acceptable range (40 degrees and rising).

- Stone: these surfaces generally provide good sealant adhesion. However, some materials such as granite, limestone, and marble should be primed before a sealant is applied. If the surface area of the stone appears to be flaking or dusting, it must be cleaned by water blasting, sandblasting, or wire-brushing before primer and sealant application.

- Glass and porcelain surfaces: these surfaces are excellent substrates for sealants once their surfaces are cleaned of contaminants, oils, and grease. Alcohol is an excellent cleaner.

- Painted and lacquered surfaces: depending upon where these surfaces are located (exterior or interior) and their exposure to weather, sealants should not be applied to flaking painted or lacquered surfaces. Sound surfaces should first be cleaned by wiping with a solvent to remove oil and dust. It is preferable to test a section to ensure that the solvent does not "lift" the painted or lacquered surface.

- Rigid plastic materials: solvents will clean these surfaces adequately, but the manufacturer of the fiberglass, acrylic, resin-based, or other plastic compound should be consulted to determine which solvent is appropriate and will not damage the plastic surface.

- Flexible plastics and elastomers: these materials are difficult to adhere to. Test applications of a solvent should be applied to determine if it cleans the product but does not damage the plastic or elastomer.

- Aluminum with a mill finish: a good degreasing material such as trichloroethylene or xylene will clean these surfaces properly. A prior rubdown with fine steel wool or fine emery on slightly oxidized surfaces might permit better adhesion.

- Aluminum with an anodized finish: this surface is generally excellent for sealant application. However, it should be wiped down with a solvent to remove any surface contamination.

- Copper: copper can oxidize to a brown or greenish patina, and this surface must be removed by either sanding or rubbing with steel wool. Copper is not compatible with many sealants, and the sealant manufacturer should be consulted for proper cleaning products.

- Lead: though not used extensively in new construction any longer, lead is often encountered in restoration work. It is difficult to obtain adhesion to a lead surface, even after a thorough cleaning with solvent such as xylene or methyl ethyl ketone. Seek advice from a sealant manufacturer.

- Steel: most steel surfaces can be readily caulked, whether painted or unpainted. On unpainted steel, all rust, oils, and other surface contaminants must be removed. Abrade the surface slightly with sandpaper, steel wool, or wire brush; clean with a solvent; dry and caulk.

- Stainless steel: this material is difficult to seal because of adhesion problems. Primers are often required after solvent cleaning.

- Galvanized steel: new galvanized surfaces present greater adhesion problems than weathered galvanized surfaces. Once again, consult a sealant manufacturer for the proper sealant material.

13

HEATING AND COOLING SYSTEMS

Heating and air conditioning are largely standard equipment in homes today. While there are still houses where air conditioning is not needed or is not present, there are very few homes without heat. Most houses have central heating systems, though some are still being heated by wood stoves and space heaters. Heating and air conditioning, for the most part, are considered to be mandatory equipment in a home.

What types of systems will you offer your customers? Many builders find heat pumps, which provide heating and air conditioning from the same unit, to be the most popular choice. This is often the case. When I was building in Virginia, all my houses had heat pumps. But when I moved to Maine, the heating system of choice was hot-water-baseboard heat. The selection for heating and cooling systems can vary from region to region.

How much do you need to know about climate-control systems? You don't have to know how to install them, but you should make yourself knowledgeable enough to discuss them adequately with your customers. If you recommend electric heat to a customer in Maine, you are going to meet a lot of resistance. The same is likely to be true if you suggest an oil-fired boiler to someone in Florida.

HEATING OPTIONS

What are your heating options? They are numerous. But, depending on where you are building, there will be some types of systems that are much more popular than others in your area. I have been working construction for over 25 years and have seen some changes in heating systems. Many basics have remained about the same, but technology has come a long way since I built my first house.

Heat Pumps

If you work in a region where temperatures are not too extreme, heat pumps are an excellent option. This type of equipment provides both heating and cooling from a single unit. The price of this combination system is extremely affordable.

Heat pumps can be used in regions where temperatures go to extremes, but the typical type of heat pump is expensive to operate when outside temperatures fall into the extremely cold range. One solution is to use a water-based heat pump. These units work well at all times, but they are costly. Most heat pumps are air-based. These are much less expensive, but when temperatures drop into frigid zones, the heat pump uses electric coils to produce heat. The operating cost for this can be very expensive over time.

The size and access requirements of a heat pump must be considered when planning construction. The outdoor unit will require a concrete pad and adequate access. Check with your subcontractor to confirm what the space requirements will be for your job.

Forced-Hot Air

Forced-hot-air heating systems are common. These are good and efficient systems. However, some people suffer from allergies and dislike the results of forced hot air. This is a question worth investigating before building a new home for someone. And don't think that you can use a forced-hot-air furnace and an independent air-conditioning unit on the same ductwork. At first thought, this seems like a viable idea. But heating systems and air-conditioning systems have different requirements for the sizing of ductwork. If you design the system for air conditioning, the heating system will not work efficiently. The same is true in reverse. This is not the case with a heat pump, but don't connect an independent heat system and an independent air conditioner to the same ductwork.

TABLE 13.1 Evaluating various types of heat pumps.

Profile of an air-source heat pump

Suitability	Good
Stability	Extreme
Availability	Excellent
Initial cost	Low
Operating cost	High
Drawbacks	Frosting

Profile of an earth-source heat pump

Suitability	Good
Stability	Stable
Availability	Excellent
Initial cost	Mid-range
Operating cost	Low to medium
Drawbacks	Leaks are hard to find and expensive to repair

Profile of a surface-water-source heat pump

Suitability	Varies
Stability	Fair
Availability	Limited
Initial cost	Mid-range
Operating cost	Low
Drawbacks	Corrosion and dry spells

Profile of a well-water-source heat pump

Suitability	Excellent
Stability	Stable
Availability	Very good
Initial cost	Mid-range
Operating cost	Low
Drawbacks	Mineral build-ups

(continued on next page)

Hot-Water-Baseboard Heat

Hot-water-baseboard heat is extremely popular in states where winter temperatures are extremely cold. These systems work with a boiler and provide even, radiant heat that can combat the coldest temperatures. When this type of heating system is used, it is rare to find a central air-conditioning system. Most homes in cold climates have mild enough summers that central air conditioning is either not needed or not justified due to the expense.

TABLE 13.1 Evaluating various types of heat pumps. *(continued)*

Profile of a solar-source heat pump

Suitability	Good
Stability	Extreme
Availability	Excellent
Initial cost	Mid-range to high
Operating cost	Low
Drawbacks	Complicated system with expensive set-up costs

Advantages of water-source heat pumps

No heating costs
Simplicity in design
Low maintenance costs
Longevity

Disadvantages of water-source heat pumps

Potential for well or surface water drying up
Mineral build-ups
Corrosion
Possible water pump failure

Types of solar-powered heat-pump systems

Solar-powered heat pumps
Passive solar heat pumps
Solar-assisted heat pumps
Solar evaporator coils

Advantages of solar evaporator coils

Provides a high source of evaporator heat
Provides a high COP from the heat pump
Minimum auxiliary heat required

Disadvantages of solar evaporator coils

Requires sunlight for best results
Minimum auxiliary heat required

Advantages of passive solar systems

No moving parts to break or wear out
Low cost to operate
Few defrost cycles

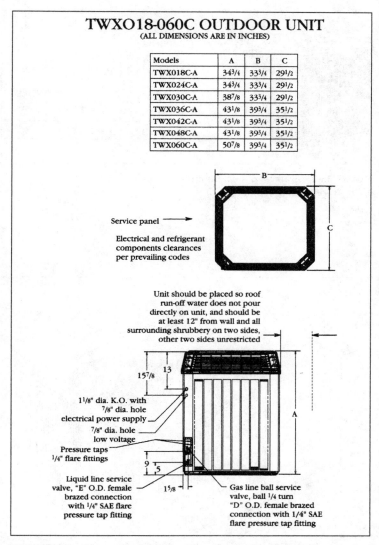

TWX018-060C OUTDOOR UNIT
(ALL DIMENSIONS ARE IN INCHES)

Models	A	B	C
TWX018C-A	34³/₄	33³/₄	29¹/₂
TWX024C-A	34³/₄	33³/₄	29¹/₂
TWX030C-A	38⁷/₈	33³/₄	29¹/₂
TWX036C-A	43¹/₈	39³/₄	35¹/₂
TWX042C-A	43¹/₈	39³/₄	35¹/₂
TWX048C-A	43¹/₈	39³/₄	35¹/₂
TWX060C-A	50⁷/₈	39³/₄	35¹/₂

Service panel

Electrical and refrigerant
components clearances
per prevailing codes

Unit should be placed so roof
run-off water does not pour
directly on unit, and should be
at least 12" from wall and all
surrounding shrubbery on two sides,
other two sides unrestricted

1¹/₈" dia. K.O. with
⁷/₈" dia. hole
electrical power supply
⁷/₈" dia. hole
low voltage
Pressure taps
¹/₄" flare fittings

Liquid line service
valve, "E" O.D. female
brazed connection
with ¹/₄" SAE flare
pressure tap fitting

Gas line ball service
valve, ball ¹/₄ turn
"D" O.D. female brazed
connection with 1/4" SAE
flare pressure tap fitting

FIGURE 13.1 Dimensional data for an outdoor unit. *(Courtesy Trane Company and American Standard Company)*

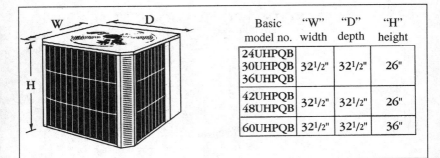

FIGURE 13.2 Example of heat pump dimensions. *(Courtesy Bard Manufacturing Company)*

Basic model no.	"W" width	"D" depth	"H" height
24UHPQB 30UHPQB 36UHPQB	32¹/₂"	32¹/₂"	26"
42UHPQB 48UHPQB	32¹/₂"	32¹/₂"	26"
60UHPQB	32¹/₂"	32¹/₂"	36"

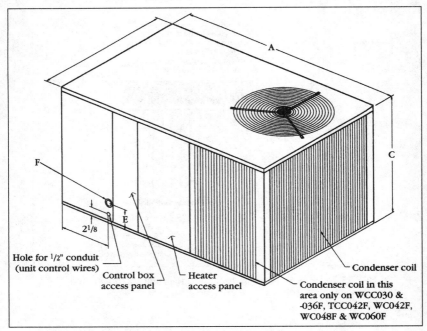

Hole for ¹/₂" conduit (unit control wires)

Control box access panel

Heater access panel

Condenser coil

Condenser coil in this area only on WCC030 & -036F, TCC042F, WC042F, WC048F & WC060F

FIGURE 13.3 Access requirements for a heat pump. *(Courtesy Trane Company and American Standard Company)*

TABLE 13.2 Conversion table for residential ductwork for 4-inch to 8-inch round branch ducts.

*CFM	Round duct size	Rectangular duct size
50	4 inch	4" × 4"
75	5 inch	4" × 5" & 4" × 6"
100	6 inch	4" × 8" & 5" × 6"
125	6 inch	4" × 8", 5" × 6", & 6" × 6"
150	7 inch	4" × 10", 5" × 8", & 6" × 6"
175	7 inch	5" × 10", 6" × 8", 4" × 14", & 7" × 7"
200	8 inch	5" × 10", 6" × 8", 4" × 14", & 7" × 7"
225	8 inch	5" × 12", 7" × 8", & 6" × 10"

*CFM = cubic feet per minute

Radiant Floor Heat

One of the most popular, fastest-growing types of heat is radiant floor heat. This type of heating system has been around for quite a while, but the options have expanded over the past several years. There was a time when this type of heat was considered to be useful only with concrete floors. This is no longer the case. There are now systems that can be installed in wood framing to provide warmth that is generated below the floor. Many builders install radiant floor heat in bathrooms and kitchens, where carpet is usually not present. A lot of new homes are heated entirely with in-floor systems. These systems require a boiler to operate.

TABLE 13.3 Conversion table for residential ductwork for 10-inch to 14-inch round main or trunk ducts.

*CFM	Round duct size	Rectangular duct size
400	10 inch	4" × 20", 7" × 10", 6" × 12", & 8" × 9"
450	10 inch	5" × 20", 6" × 16", 9" × 10", & 8" × 12"
500	10 inch	10" × 10", 6" × 18", 8" × 12", & 7" × 14"
600	12 inch	6" × 20", 7" × 18", 8" × 16", & 10" × 12"
800	12 inch	8" × 18", 9" × 15", 10" × 14", & 12" × 12"
1000	14 inch	10" × 18", 12" × 14", & 8" × 24"

*CFM = cubic feet per minute

TABLE 13.4 Conversion table for residential ductwork for 16-inch to 20-inch round main or trunk ducts.

*CFM	Round duct size	Rectangular duct size
1200	16 inch	10" × 20", 12" × 18", & 14" × 15"
1400	16 inch	10" × 25", 12" × 20", 14" × 18", & 15" × 16"
1600	18 inch	10" × 30", 15" × 18", & 14" × 20"
1800	20 inch	10" × 35", 15" × 20", 16" × 19", 12" × 30", & 14" × 25"
2000	20 inch	10" × 40", 12" × 30", 15" × 25", & 18" × 20"

*CFM = cubic feet per minute

Electric Heat

Electric baseboard heat is very inexpensive to buy and install. The expense with this type of heating system comes in the operating costs. Electric heat is a good choice for homes in warm regions where heat is rarely needed. Another good use of electric heat is in the role of back-up heat. If you have a customer who plans to heat a home mostly with a wood stove, electric heat makes an ideal back-up heat.

Space Heating

Space heating, whether with a wood stove or a direct-vent heater of some other type, is not going to satisfy most lenders as a suitable heating system. It is common for mortgage lenders to require a central heating system. If you have a customer who is committed to

TABLE 13.5 Conversion table for residential ductwork for 9-inch to 12-inch round branch ducts.

*CFM	Round duct size	Rectangular duct size
250	9 inch	6" × 10", 8" × 8", & 4" × 16"
275	9 inch	4" × 20", 8" × 8", 7" × 10", 5" × 15", & 6" × 12"
300	10 inch	6" × 14", 8" × 10", 7" × 12"
350	10 inch	5" × 20", 6" × 16", & 9" × 10"
400	12 inch	6" × 18", 10" × 10", & 9" × 12"
450	12 inch	6" × 20", 8" × 14", 9" × 12", & 10" × 11"

*CFM = cubic feet per minute

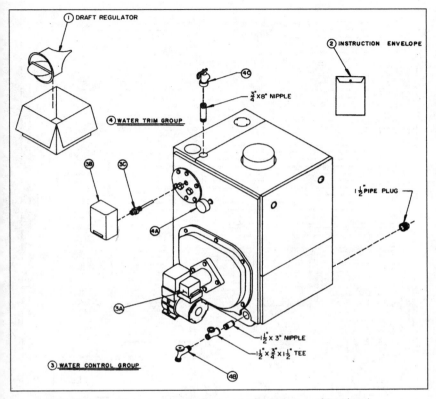

① DRAFT REGULATOR

② INSTRUCTION ENVELOPE

④C

$\frac{3}{4}$" X8" NIPPLE

④ WATER TRIM GROUP

③B ③C

1 $\frac{1}{2}$" PIPE PLUG

④A

③A

1 $\frac{1}{2}$" X 3" NIPPLE

1 $\frac{1}{2}$" X $\frac{3}{4}$" X 1 $\frac{1}{2}$" TEE

③ WATER CONTROL GROUP

④B

FIGURE 13.4 Water boiler trim and controls. *(Courtesy of Burnham)*

using a space heater, consider installing electric baseboard heating units to meet bank requirements.

AIR CONDITIONING

Air conditioner options come down to two basic choices. You can offer a heat pump that provides heating and cooling from a single system, or you can use a single heating system and a separate cooling system. The heat pump is the most cost-effective option in most regions. A separate cooling system adds quite a bit to the combined cost of a heating and cooling package.

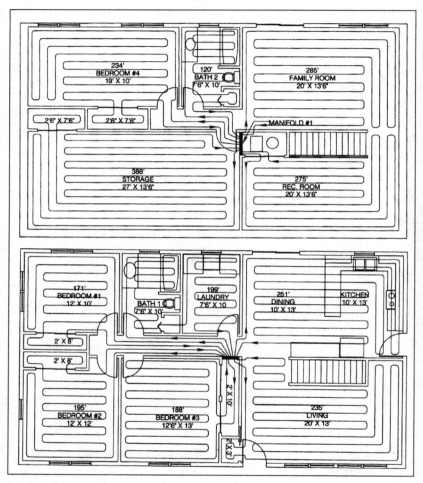

FIGURE 13.5 Examples of plans for in-floor heating systems. *(Courtesy of Wirsbo)*

ROUGH-IN

The rough-in for a heating and/or cooling system will vary, depending on the type of system being installed. Roughing-in for electric heat is simply a matter of having your electrician run wires to the desired locations and leaving them for use in the final phase. If you

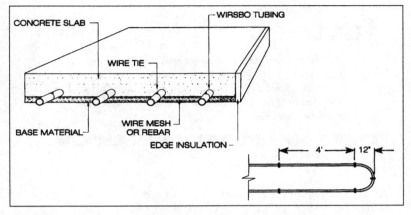

FIGURE 13.6 Radiant floor heating in a slab on grade or a slab below grade. *(Courtesy of Wirsbo)*

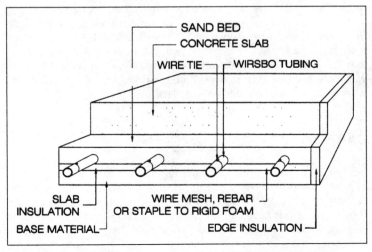

FIGURE 13.7 Slab installation over a sand bed. *(Courtesy of Wirsbo)*

fastfacts

When installing a heating or cooling system that requires duct-work, you have to plan the installation locations well in advance. There may be a need to build chases, head off joists, rearrange plumbing, or any number of other compromises to accommodate the ducts.

are installing an in-floor system, you will have to let your plumbing or heating contractor route the piping that will be concealed in the flooring. This piping will need to be tested for leaks prior to concealment. If the piping is located in a place where it might be damaged by a nail or screw, nail plates will be needed to protect the piping. Most jurisdictions will require an inspection of any installation that will be concealed prior to the concealment.

A rough-in for a hot-water-baseboard system consists of ¾ inch copper tubing being stubbed up through the floor. Depending upon the house style, the horizontal piping might be attached to the bottom of floor joists. If the joists below the finished floor will have a finished ceiling, such as in a two-story house, the feed and return pipes will have to be installed within the joist bays.

When air conditioning or a heat pump is installed, there will be ductwork to deal with. This is also true with a forced-hot-air heating system. You will have to plan the installation of the ductwork carefully to avoid conflicts with carpentry and plumbing work. Sit down with your installer and go over the blueprints carefully to determine where major ductwork will be installed.

CARPENTERS

It is good thinking to have a carpenter on any job when ductwork is being installed. It is not uncommon for the installers to require the cutting of joists. When this is the case, the joists will have to be headed off. If you have carpenters on the job, the duct installers and carpenters can discuss the best strategies before cutting wood members. While this advice is not essential to a job, it can make the job easier.

fastfacts

When using a heating system that requires ductwork, send your installers in before you send in your plumbers or electricians. Otherwise you may have to relocate plumbing or wiring to make room for ducts. Even if you have the job planned perfectly, things can go wrong, so send in the duct installers first.

CHOOSING EQUIPMENT

Choosing equipment after you have decided on a type of system can be confusing. There are many choices available, and your customers are likely to become confused. Unlike plumbing, where a toilet operates pretty much the same regardless of the brand, this is not the case with heating and cooling systems. Picking light fixtures is mostly a matter of taste in physical appearance. The same is basically true for most plumbing fixtures. But, when you get into heating systems, there are far more factors to consider. Consider some of the following questions:

- Will the equipment be fueled by electricity?
- Will the equipment be fueled by natural gas?
- Will the equipment be fueled by LP gas?
- Will the system be fueled by oil?
- If a boiler is used, will it be a steel boiler or a cast-iron boiler?
- Will a cold-start boiler be used?
- What brand of equipment offers the best warranty for your customer?

fastfacts

Don't conceal any rough-in work before it is inspected and approved by the local code-enforcement official.

- Is there a particular brand of equipment that is most desirable in your region?

Becoming knowledgeable about heating and cooling systems may be your biggest responsibility in this phase of work. It will take some time to go over fact sheets and specifications, but they are readily available from manufacturers, so you won't have any trouble finding the data to assess. Dig into it a little and explore the options that you can offer your customers. Going this extra distance can help to set you apart from average builders.

EXTERIOR WORK

Depending on the types of systems you will be installing, you may have some exterior work to consider. While it is true that your sub-contractor will be responsible for the actual work, you still have some thinking to do as the builder. For example, let's assume that you are having a typical heat pump installed. This will require an outside unit, and it will need a pad to sit on. Where will you have it installed? Have you considered the appearance of the unit? Think about it and discuss it with your customers. When you will be installing exterior units, you must factor in the unit and pad with your final grading and landscaping plans.

FINISH WORK

The finish work with heating and cooling systems does not require a lot of effort on a builder's part. If you are having a ducted system installed, you should check all the register grates to see that they fit up properly and work as they should. With baseboard heating systems check to make sure that all end caps and trim are installed properly. Every element of the system should be checked carefully before turning a home over to a customer.

14

PLUMBING

Plumbing is present in nearly all new construction. This is a phase of work that makes many builders nervous. Good plumbers can be hard to find, but the work that they do is not extremely complicated. As a builder, you don't need to understand all of the plumbing code or how to solder joints. It is helpful to have enough knowledge to check rough-in measurements. If a plumber installs rough plumbing in the wrong place and you don't find out about it until you are ready to have the plumbing fixtures set, you are going to be in for a mess.

Plumbers normally begin work on a job after the HVAC mechanics and before the electricians. There is no rule that says this must be the case, but it is the most logical approach. Why? HVAC installers have little flexibility on where to install their large ducts. If plumbing is in the way of where a trunk line must go, the plumbing has to be moved. Plumbing offers less flexibility in routing that electrical wiring does. This is why the plumbing follows the ductwork and precedes the electrical work. Keep this in mind when working out your production schedule.

While you don't need to understand all the elements of plumbing work, it is helpful to have some feel for what will happen on your jobs. For this reason, we will look at some of the high spots for typical plumbing work.

fastfacts

When you are scheduling your mechanical trades, you should normally schedule your HVAC work first, your plumbing second, and your electrical work last. This is generally the most productive way in which to install the mechanical systems.

WATER DISTRIBUTION

Installing water-distribution systems is usually easier than installing a drain-waste-and-vent (DWV) system. For one thing, the size of the pipes being installed with a water system is smaller. The requirements for pipe grading are also less stringent. Code requirements for water systems are easier for most plumbers to understand. When all elements are considered, a residential water system is quite simple. Commercial jobs are more complex, but piping diagrams are supplied that make the installations easy if you are good at following directions.

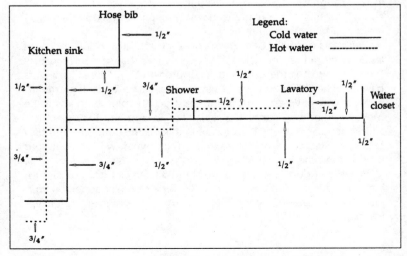

FIGURE 14.1 Riser diagram illustrating efficient water-supply design.

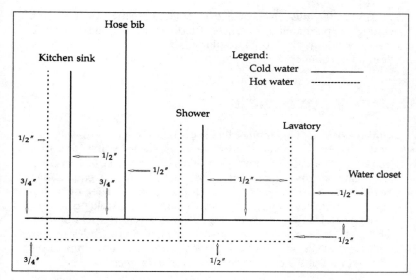

FIGURE 14.2 Riser diagram illustrating inefficient water-supply design.

POTABLE WATER

Potable water is required for drinking, cooking, bathing, and the preparation of food and medicine. Potable water is also used for other activities, but the ones listed above are the uses dealt with in the plumbing code. Fixtures used for any of the above purposes must be plumbed in a way that only potable water is accessible to them.

fastfacts

- *Potable water is required for cooking.*
- *Potable water is required for drinking.*
- *Potable water is required for the preparation of food.*
- *Potable water is required for the preparation of medicine.*
- *Potable water is required for bathing.*

Hot water must be provided in buildings where people work and for all permanent residences. Cold water must be supplied to every building that contains plumbing fixtures and is used for human occupancy.

ANTI-SCALD DEVICES

Anti-scald devices are required on some plumbing fixtures. Different codes require these valves on different types of fixtures, and they sometimes set different limits for the maximum water temperature. Anti-scald devices are valves or faucets that are specially equipped to avoid burns from hot water. These devices come in different configurations, but they all have the same goal: to avoid scalding.

Whenever a gang shower is installed, such as in a school gym, anti-scald shower valves should be installed. Some codes require anti-scald valves on residential showers. The maximum temperature allowed for hot water in some codes is 110 degrees F.

CHOOSING PIPES AND FITTINGS

The three primary types of pipe used in above-grade water systems are cross-linked polyethylene (PEX), copper, and CPVC. All of these pipes will provide adequate service as water distribution carriers. The pipe you choose will be largely a matter of personal preference. Copper pipe and tubing has long been the standard for professional plumbers, but PEX is being used more and more often.

WATER SERVICE

Is a water service part of a water-distribution system? It is part of the water system, but it is not normally considered to be a part of a water-distribution system. This can be confusing, so let me explain.

A water service does convey potable water to a building, but it is not considered a water-distribution pipe. Water services fall into a category of their own. Water-distribution pipes are the pipes found inside a building. Water services run the majority of their length outside a building's foundation. They are usually buried in the ground. Water-distribution pipes are installed within the foundation of the building and don't normally run underground or outside a foundation. However, they can, as in the case of a one story-building where the water-distribution pipes are buried under a concrete floor. In any event, water services and water-distribution pipes are dealt with differently by the plumbing codes.

Why are water-distribution pipes considered separately from water services? Because they serve different needs. Water-distribution pipes pick up where water services leave off. The pipes distributing water to plumbing fixtures do not have to meet the same challenges that underground water services do.

For one thing, the water pressure on a water-distribution pipe is often less than that of a water service. Most homes have water pressure ranges between 40 psi and 60 psi. A water service from a city main could easily have an internal pressure of 80 psi or more.

Water services are also buried in the ground. Water-distribution pipes rarely are. The underground installation can affect the types of pipes that are approved for use. Another big difference between water service pipes and water-distribution pipes is the temperature of the water they contain. Water services deliver cold water to a building. Water-distribution pipes normally distribute both hot and cold water. Since most codes require the same type of pipe to be used for the cold water lines as is used for hot water pipes, the temperature ranges can disqualify some types of materials. This is a significant factor when choosing a water-distribution material.

BACKFLOW PROTECTION

Backflow protection for water systems has become a major issue in the plumbing trade. The forms of protection required cover such issues as cross-connections, backflow, and vacuums. If backflow preventers are not installed, entire water mains and water supplies can be at risk of contamination.

Backflow preventers are devices that prohibit the contents of a pipe from flowing in a direction opposite its intended flow. These protection devices come in many shapes and sizes, and costs range from a few dollars to thousands of dollars. Backflow preventers must be installed in accessible locations.

When a potable water pipe is connected to a boiler to provide it with a supply of water, the potable water pipe must be equipped with a backflow preventer. If the boiler has chemicals in it, such as anti-freeze, the connection must be made through a reduced-pressure-principal backflow preventer or an air gap.

Whenever a connection is subject to back pressure, the connection must be protected with a reduced-pressure-principal backflow preventer.

AIR GAPS

An air gap is the open air space between the flood level rim of a fixture and the bottom of a potable water outlet. For example, there is an air gap between the outlet of a spout on a bathtub and the flood level rim of the tub, or at least there is supposed to be. By having such an air gap, non-potable water, such as dirty bath water, cannot be sucked back into the plumbing system through the potable water outlet. If the tub spout was mounted in a way that it was submerged in the tub water, the dirty water might be pulled into the water pipes if a vacuum was formed in the plumbing system.

The amount of air gap required on various fixtures is determined by the plumbing code. For example, you might find in a table or in the text of your code that a tub spout must not be installed in a way that its opening is closer than 2 inches to the flood level rim. The same ruling for lavatories might require the spout of the faucet to remain at least 1 inch above the flood level rim.

There is another type of air gap used in plumbing. This type is a fitting that is used to accept the waste from a dishwasher for conveyance to the sanitary drainage system. The device allows the water draining out of the dishwasher to pass through open air space on its way to the drainage system. Since the water passes through open air, it is impossible for wastewater from the drainage system to be siphoned back into the dishwasher.

VACUUM BREAKERS

Vacuum breakers are installed to prevent a vacuum being formed in a plumbing system that could result in contamination. Some of these devices mount in-line on water distribution pipes and others screw onto the threads of silcocks and laundry-tray faucets.

When vacuum breakers are required, the critical level of the device is usually required to be set at least 6 inches above the flood level rim of the fixture it is serving. What is the critical level? The critical level is the point at which the vacuum breaker would be submerged prior to backflow.

Normally, any type of fixture that is equipped with threads for a garden hose must be fitted with a vacuum breaker. Some silcocks are available with integral vacuum breakers. For the ones that are not, small vacuum breakers that screw onto the hose threads are available. After the devices are secured on the threads, a set screw is turned into the side of the device. The screw is tightened until its head breaks off. This ensures that the safety device cannot be removed easily.

PENETRATING STUDS AND JOISTS

Water pipes are normally installed in the floor or ceiling joists so they can be hidden from plain view. When possible, keep the pipe at least 2 inches from the top or the bottom of the joists. Keep in mind that your local building code will prohibit you from drilling, cutting, or notching joists close to their edges. Keeping pipes near the middle of studs and joists reduces the risk of the pipes being punctured by nails or screws. If the pipe passes through a stud or joist where it might be punctured, protect it with a nail plate.

PIPE SUPPORT

Proper pipe support is always an issue that all plumbers must be aware of. Copper tubing with a diameter of 1¼ inches or less should be supported at intervals not to exceed 6 feet. Larger copper can run 10 feet between supports. These recommendations are for pipes installed horizontally. If the pipes are run vertically, both sizes should be supported at intervals of no more than 10 feet.

fastfacts

After plumbing is installed and before you allow the walls and ceilings to be covered, check the job to see that all the piping is installed in such a way as to avoid nails or screws from hitting the pipes.

fastfacts

Avoid placing water pipes in outside walls and areas that will not be heated, such as garages and attics.

If your material of choice is CPVC, it should be supported every 3 feet, whether installed vertically or horizontally. PEX pipe that is installed vertically is required to be supported at intervals of 4 feet. When installed horizontally, PEX pipe should have a support every 32 inches.

Rough-in Numbers

Refer to a rough-in book (available from plumbing suppliers for various brands of fixtures) to establish the proper location for your water pipes to serve the fixtures. The examples given here are common rough-in numbers, but your fixtures may require a different rough-in. The water supplies for sinks and lavatories are usually placed 21 inches above the sub-floor. Kitchen water pipes are normally set 8 inches apart, with the center point being the center of the sink. Lavatory supplies are set 4 inches apart, with the center being the center of the lavatory bowl.

Most toilet supplies will rough-in at 6 inches above the sub-floor and 6 inches to the left of the closet flange, as you face the back wall. Showerheads are routinely placed 6 feet 6 inches above the sub-floor. They should be centered above the bathing unit. Tub spouts are centered over the tub's drain and roughed-in 4 inches above the tub's flood level rim. The faucet for a bathtub is commonly set 12 inches above the tub and centered on the tub's drain. Shower faucets are generally set 4 feet above the sub-floor and centered in the shower wall. If you go to your supplier and ask for a rough-in book, the supplier should be able to give you exact information for roughing-in your fixtures.

MANIFOLD SYSTEMS

The rules for plumbing a water distribution system are basically the same for all approved materials. But when running PEX it often

fastfacts

- *Water service pipe must have a minimum pressure rating of 160 psi at 73.4 degrees F.*
- *The bottom of a water service pipe must be at least 12 inches above the top of a sewer.*
- *When a water service is installed in a trench with a sewer, the water service pipe must be installed on a solid shelf of firm material to one side of the trench.*
- *A water service pipe must be protected from freezing temperatures.*
- *Water service pipes should be protected from abrasive surfaces.*

makes more sense to use a manifold than to run a main with many branches. One of the most desirable effects of this type of installation is the lack of concealed joints. All the concealed piping is run in solid lengths without fittings.

A manifold can be purchased from a plumbing supplier. It will come with cut-off valves for the hot and cold water pipes. Water comes into one end of the manifold and goes out the other. The manifold is divided into hot and cold sections. The pipe from each fixture connects to the manifold. By installing a water distribution system in this way, you save many hours of labor. You also eliminate concealed joints and can cut off any fixture with the valves on the manifold. This type of system is efficient, economical, and the way of the future.

INSTALLING DRAIN-WASTE-AND-VENT (DWV) SYSTEMS

There is no question that DWV systems require planning. It's usually easy to change directions with a water pipe, but this is not always the case with a drain or vent. Thinking ahead is crucial to the success of a job. This is true of any type of plumbing, and the rule certainly applies to DWV systems.

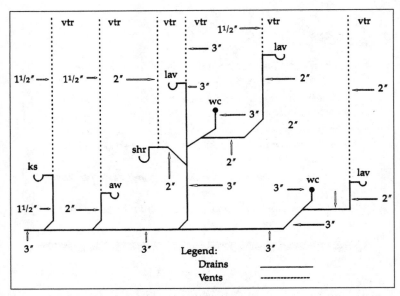

FIGURE 14.3 Poorly designed DWV layout.

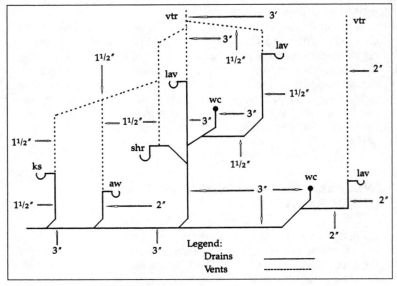

FIGURE 14.4 Efficient use of DWV pipes.

UNDERGROUND PLUMBING

Underground plumbing is installed before a house is built. When the plumbing is not installed accurately, you will have to break up a new concrete floor to relocate the pipes. Since the placement of underground plumbing is critical, you must be acutely aware of how it is installed. Miscalculating a measurement in the above-grade plumbing may mean cutting out the plumbing, but the same mistake with underground plumbing will be much more troublesome to correct.

An underground plumbing system is often called groundwork among professional plumbers. The groundwork is routinely installed after the footings for a home are poured and before the concrete slab is installed. After your plumbing contractor installs underground plumbing, check the rough-in measurements. The time you spend doing this can save you a lot of trouble as the project moves forward.

ABOVE-GROUND ROUGH-INS

Once a building is framed and under roof, you can start the installation of above-ground drains and vents. If the building has underground

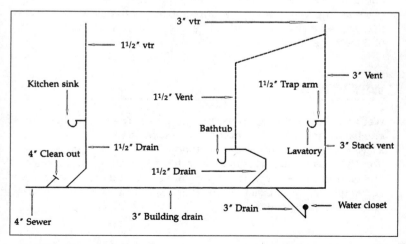

FIGURE 14.5 DWV riser diagram, with size and location of pipes.

fastfacts

Generally the grade is set at ¼ inch per foot for drains and vents. Drains fall downward toward the final destination. Vents pitch upward toward the roof of the house. With a 12-foot piece of pipe, the low end will be 3 inches lower than the high end. Avoid having either too much or too little grade. You can purchase a grade level to check behind your plumbers if you want to.

plumbing, your plumber will be tying into the pipes you installed earlier. Not all houses have ground works, so the framing stage might be the plumber's first visit to a job for installation.

Hole Sizes

Hole sizes are important. Some codes require you to keep the hole size to a minimum to reduce the effect of fire spreading through a home. If oversized holes are used, they can act as a chimney for fires in a building. Make sure that your plumbing contractor understands that you want any holes made to be as small as possible. In the case of 2-inch pipe, a standard drill bit size will be 2⁹⁄₁₆ inches.

WET VENTS

Wet vents are pipes that serve two purposes. They are a drain for one fixture and a vent for another. Toilets are often wet-vented with a lavatory. This involves placing a fitting within a prescribed distance from the toilet that serves as a drain for the lavatory. As the drain proceeds to the lavatory, it will turn into a dry vent after it extends above the trap arm. Exact distances and specifications are set forth in local plumbing codes.

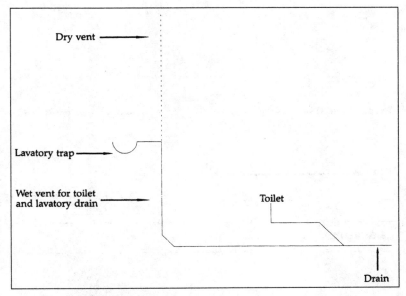

FIGURE 14.6 Example of a wet vent.

DRY VENTS

Many fixtures will be vented with dry vents. These are vents that do not receive the drainage discharge of a fixture. Since the pipes only carry air, they are called dry vents. There are many types of dry vents; they include: common vents, individual vents, circuit vents, vent stacks, and relief vents, to name a few. Don't let this myriad of vents intimidate you. When plumbing an average house, venting the drains does not have to be complicated.

FIXTURE PLACEMENT

Fixture placement is normally shown on all blueprints. There are code requirements that dictate elements of locating fixtures.

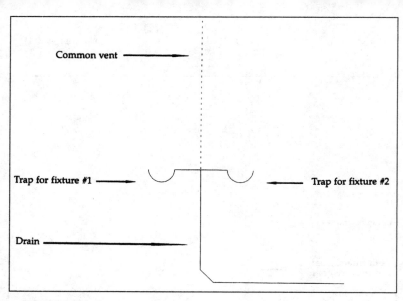

FIGURE 14.7 Common vent.

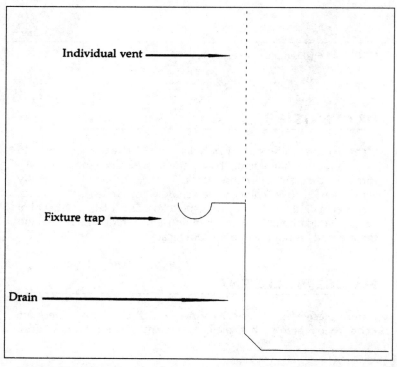

FIGURE 14.8 Individual vent.

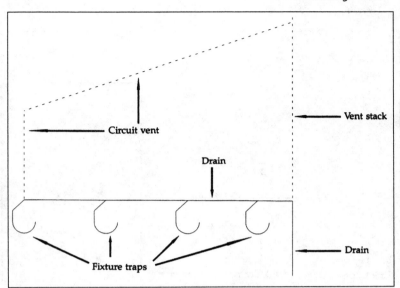

FIGURE 14.9 Circuit vent.

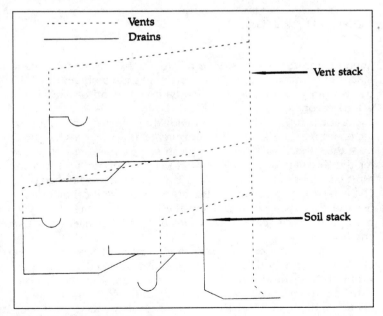

FIGURE 14.10 Relief vent.

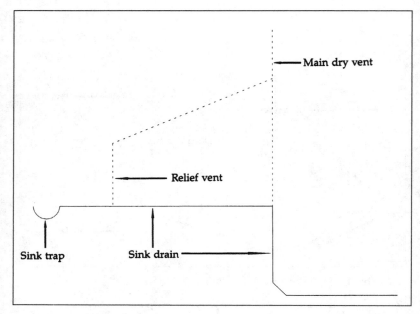

FIGURE 14.11 Vent stack.

Rough-in measurements vary from fixture to fixture and from manufacturer to manufacturer. To be safe, you should obtain rough-in books from your supplier for each type and brand of fixture you will be roughing in.

Standard toilets will rough in with the center of the drain 12 inches from the back wall. This measurement of 12 inches is figured from the finished wall. If you are measuring from a stud wall, allow for the thickness of drywall or whatever the finished wall will be. Most plumbers rough-in the toilet 12½ inches from the finished wall. The extra ½ inch gives you a little breathing room if conditions are not exactly as you planned. From the front rim of the toilet, you must have a clear space of 18 inches between the toilet's rim and another fixture.

From the center of the toilet's drain, you must have 15 inches of clearance on both sides. This means you need a minimum width of 30 inches to install a toilet. If you are plumbing a half-bath, the room must have a minimum width of 30 inches and a minimum depth of 5 feet These same measurements for clearance apply to bidets.

fastfacts

If you plan to install a tub waste using slip-nuts, you will have to have an access panel to gain access to the tub waste. If you do not want to have an access panel, use a tub waste with sol-vent-weld joints. Many people object to access panels in their hall or bedroom; avoid slip-nut connections and you can avoid access panels.

FINAL PLUMBING

When you get to the final plumbing, check all the fixtures carefully for defects. Look for leaks. The local plumbing inspector will do this during the required final inspection, but it never hurts to look for yourself before turning a house over to a customer. If you have a good plumber, your effort as a builder in the plumbing phase will be minimal, but it's your job, so keep an eye on it.

15

ELECTRICAL SYSTEMS

Electrical work is, in many ways, very similar to plumbing. It's not that the work itself is the same, but the risks that each phase presents to a general contractor are closely related. Builders are constantly faced with electrical installations. Having a general idea of what to expect is very helpful when running jobs.

Electricians are required to be licensed in order to practice their trade. This licensing requirement prohibits average people, like most contractors, from doing their own electrical wiring. This puts pressure on you to find good subcontractors in the electrical field. How can you tell if your potential electrician is any good? References from other jobs help, and the fact that the trade is licensed doesn't hurt. But, the best way to tell is to know something about the trade. I feel strongly that it pays to put yourself in the shoes of each trade you will be dealing with. If you don't, telling the difference between a good subcontractor and a bad one is much more difficult.

ELECTRICAL CONCERNS

Electrical concerns are likely to come up in just about any discussion of a major building project. Customers will want to know what their electrical needs will be. Some of the most important questions may never be asked. Why is this? Simply because consumers are not tuned into electrical codes, so they don't know what questions to ask. They may assume many things that are not correct. It is your job, and that of your electrician, to educate these people.

What Are These Ground-Fault Things?

A typical question for a homeowner might be, "What are these ground-fault things that I've heard about?" People have heard about ground-fault interceptors (GFI), but most of the folks don't know what these devices are, when they are needed, or how they work. I assume that you are aware of what GFI devices are, but let me expand on this subject.

GFI devices can be in the form of circuit breakers or individual outlets. The outlets are much cheaper. These devices are installed in "wet" areas. The purpose of a GFI device is to protect people from hazardous electrical shocks. Since water and electricity don't make a good match in terms of personal safety, GFI breakers or outlets are required in certain locations. These locations typically included the following:

- Bathrooms
- Kitchens
- Outside outlet locations
- Darkrooms
- Laundry Rooms
- Garages

Basically, a GFI protection device should be installed in any location where a person may be in contact with water and electricity at the same time. For example, a whirlpool bathtub should be wired into a GFI-protected circuit. This would be done with a GFI circuit

fastfacts

Breakers can be used for all GFI circuit locations, but GFI outlets are a less expensive option. By making the first outlet in the circuit a GFI outlet, all remaining outlets on the circuit are GFI-protected.

breaker. Breakers can be used for all GFI circuit locations, but GFI outlets are a less expensive option. By having the first outlet in the circuit be a GFI outlet, all remaining outlets on the circuit are GFI-protected. The electrical code requires GFI protection in locations like those just mentioned, so GFI protection can't be considered an option; it is a necessity.

THE UNASKED QUESTIONS

The unasked questions pertaining to electrical work are often more important than the ones that are presented. Most homeowners don't know enough about electrical systems to understand what questions to pose. This is where your knowledge and expertise shine. If you can point out key issues to potential customers before they are even aware that such issues exist, you have a much better chance of creating a happy customer.

Take the time to talk with your customers when planning construction. It can be a good idea to have your electrician meet with you and your customers. Many residential blueprints don't have elaborate electrical plans drawn; and when they do, average people may not understand them. Adding wiring during the rough-in phase is not complicated, but doing it once the drywall is hung is a mess.

ARE YOU AWARE OF THE CODE REQUIREMENTS?

Are you aware of the code requirements pertaining to this work you want done? This question should always be asked. Honest home-

fastfacts

Planning prior to construction is essential to a successful job. Sit down with your customers and go over as many elements of a project as you can before you give a final contract price and begin work. Don't be shy. Make suggestions to your customers. Remember that they may not know the questions to ask or the options to suggest.

fastfacts

Invest in codebooks for all trades. Read the codes and familiarize yourself with them. As your subcontractors questions if you have any. I suggest leaving direct code quotes to customers in the hands of your licensed experts, but this is not to say that you should not have a pretty good idea of what code requirements are for all the trades on your jobs.

owners will almost always say that they are not aware of code requirements. This is your chance to show off a little and to impress the homeowner. For instance, you can quote the need for GFI protection. This can be followed with code requirements on outlet spacing. Essentially, wall outlets cannot be more than 12 feet apart. The code requires a light or appliance with a 6-foot electrical cord to be placed in any location along a wall without having to use an extension cord to reach an outlet. Along kitchen counters, the spacing is reduced to 4 feet. These types of statistics impress people, and if you can impress them, you can sell them. However, don't take the code regulations that I'm giving you here for gospel;check your local code for current requirements.

HAVE YOU THOUGHT ABOUT YOUR SWITCH LOCATIONS?

Have you thought about your switch locations and which switches will control which lights? This question can be very important during the planning stage of a job. By showing people how they can specify the location of switches, within reason, and how they can suggest which lights the switches operate, you can do a better job for the customer. People pick up on these types of questions, and they identify them with caring and concern on your part. This is a big step towards closing a sale.

The electrical code mandates minimum switch locations. But customers will often appreciate having more than the minimum requirements met. Look at the convenience factor and discuss options with your customers. Remember, it is your job to build a long list of satisfied customers.

HAVE YOU SET A BUDGET?

Have you set a budget for your light fixtures? Most electricians bid their jobs without including the price of light fixtures. They will typically detail a fixture allowance in their bid, but their price may not include any allowance for fixtures. If you forget to mention this fact to your customers, someone might have to come up with hundreds of dollars that he or she hadn't planned on. One fixture, such as a dining-room chandelier, can cost several hundred dollars. It is also possible to buy fixtures for less than ten dollars. Before you can bid a job successfully, you must establish a lighting allowance with your customer. You might think that a lighting allowance of $1,000 will be generous and then find out that your customers have specialized tastes that will consume that allowance quickly and leave them without the types of lighting that they desire.

TEMPORARY POWER

Temporary power for new construction can come from electric generators or temporary power poles. The temporary poles offer more dependable service, and there is no noise from running generators. The drawbacks to temporary power services are cost and the time that it can take to have them installed. Still, most builders agree that a temporary power service is the best option for building houses and larger structures.

If you expect to use a temporary power service, you will have to get the paperwork rolling early to have power when you need it. Procedures may vary from power company to power company, but I will give you an idea of what to expect.

The first step is calling the power company and requesting the required paperwork. Have the legal description of the building site available. This normally includes a map number, a lot number, and

fastfacts

Tie down as many details as possible in writing. Don't leave room for confusion or arguments. Many good jobs have gone bad due to poor attention to detail in written contracts.

fastfacts

Arranging for temporary power service can take months at times. If you want to avoid using generators on your jobs, start the process of getting temporary power early in the planning stages of construction projects.

sometimes a block number. A form will have to be filled out and returned to the power company to begin the process. Another form will have to be completed by your electrician once the temporary pole is set, wired, and ready for inspection by the power company.

Some electricians will have temporary power poles prefabricated. They may require you to buy the pole setup, or they may rent it to you. Once a site inspection is done by the power company and a suitable location for the pole is established, your electrician can install the pole, fill out the inspection form, and notify you to call the power company for an inspection.

Once the inspection is completed, the power company will energize the pole. This process can take weeks or months, so don't put getting temporary power on a back burner.

OVERHEAD OR UNDERGROUND?

Will the structure you are planning to build have an overhead electrical service or an underground service? You may not have a choice. Some subdivisions require underground service. You will have to

fastfacts

Always check deed restrictions, covenants, and easements before committing to a job. These factors can have a major impact on your job. For example, you may be required to install an underground electrical service.

check the covenants and restrictions in property deeds to see if they will affect the type of service that will be used.

Underground services are more expensive than overhead services. However, they are often more desirable from an appearance point of view. If you have the option to use either type, ask your customers what they want. Get your electrician to price the job both ways and allow your customers to choose the type of service that fits their needs and budgets.

ROUGH-IN

If you have done your planning properly, the electrical rough-in should go smoothly. The electrician will do the work, but you should check it yourself. A code officer will normally inspect the work for code violations. What you should be looking for is to see that all the device locations and any special provisions, such as wiring for telephones or computers, have been installed. You don't want to allow the walls and ceilings to be covered until you are sure that all the rough wiring is installed as per your contract and approved by the local code officer.

FIXTURES

Light fixtures will have to be selected before the final electrical work can be done. This process is sometimes simple and sometimes difficult. As a builder, I have seen many jobs where the selection of light fixtures was time-consuming. Homebuyers occasionally have trouble finding what they want within the budget that they have. There are different schools of thought on how to accomplish the task of getting electrical fixtures on the job site. Personally, I have often gone to lighting showrooms with my customers to assist them in the selection process. My presence has seemed to make the process faster and more effective.

Some builders just give the customers a dollar amount and send them to a showroom to make their selections. This can save you some time, but it might not. You will have to decide what works best for you, but I suggest that you go with your customers. Regardless of how you decide to approach the selection process, don't wait too long to get it started. Some fixtures may have to be ordered. I have customers make their selections once the rough-in is complete.

fastfacts

- *Have customers select their light fixtures once the rough-in work is done.*
- *Consider going with your customers when they are selecting light fixtures.*
- *Inspect all fixtures immediately once they are on the job site.*
- *Use written change orders if your customers are over or under budget.*

Once the light fixtures are selected, you can either have them delivered to your job site, have your electrician pick them up, or have one of your people take them to the site. Once they are on the site, check each fixture immediately for damage. The fixtures can be fragile, and if they are broken, you will need time to get replacements without delaying the job.

FINAL WORK

Final work is always crucial. This is the work that customers will live with for a long time. There is not a lot of work needed by a builder at this stage. Your primary responsibility will be to go through the building and confirm that all fixtures are installed properly and that all devices function. The code officer will be looking for this same type of thing, but it is a good idea for you to check the job yourself before turning it over to your customers. As with any phase of construction, learn as much as you reasonably can about the work being done and supervise it to the best of your ability.

16

INSULATION

Working around insulation can be an aggravating experience. Many people find it to be very irritating to their
skin. Some contractors have trouble breathing normally
when working with it. Yet, insulation is a part of every modern construction job. This means that you will come into contact with it at
some time in your work. Not all jobs put contractors face to face
with insulation, but some do. Even if you aren't involved in the
installation of new insulation, you are very likely to have to put up
with insulation in one form or another.

Have you ever removed a ceiling and had old insulation rain
down on you? If you have, you know how uncomfortable the rest of
the workday can be. Getting insulation in your face, in your clothes,
and all over you can make finishing out the day a real struggle. Even
if you are wise enough to check above a ceiling before you open it,
you may not be able to avoid old insulation. When this is the case,
the best you can do is prepare to work around it.

Installing new insulation is not a very technical job. Almost anyone can learn quickly how to install most forms of insulation. As a
contractor, you can have one of your laborers or trainees do the
dirty work for you, but this may not prove to be profitable. It is one
thing to be able to install insulation; it is quite another to be competent enough to install it profitably.

fastfacts

After more than 25 years in the business, I have never found a way to install insulation with employees at a cost that is less than what I can pay to subcontract it to companies that specialize in the installation of insulation. Before you get itchy installing your own insulation, get some bids from insulation companies. I don't know for sure what your bids will be, but I have never been able to find a way to beat the prices of a company that concentrates solely on insulation.

Insulation is one phase of construction work that I've never enjoyed. Try as I have, there seems to be no way to escape working with it. I have used helpers to do installations, but I've never made much money working in this manner. My experience has proved that, for me at least, more money can be made by hiring specialized insulation companies to do the installation work for me. I don't know how the companies make money. I've found that aggressive insulators will provide the labor and material to insulate a job for about what I have to pay just for the insulation. This essentially means that I am getting the installation for free.

Unless you are working on an extremely tight budget and doing most of the work with your own two hands, there is little justification in doing your own insulation work. A good crew from a professional insulating company can get the job done much faster than you can, and the cost will probably not be much more than what you would spend on materials. Your time can be put to better, and more profitable, use. Go out and sell another job; that's where the real money is.

TYPES OF INSULATION

There are many types of insulation available. Some types are better suited for certain kinds of jobs than others. The R-values of insulation vary. While most contractors have standard procedures for their insulation installations, it can pay to know what all your options are. As with any other aspect of your business, the better informed you are, the better off you are.

TABLE 16.1 R-values for insulation.

Material	R-value per inch of insulation
Fiberglass batts	3
Fiberglass blankets	3.1
Fiberglass loose-fill	3.1 to 3.3 (when poured), 2.8 to 3.8 (when blown)
Rock-wool batts	3
Rock-wool blankets	3
Rock-wool loose-fill	3 to 3.3 (when poured), 2.8 to 3.8 (when blown)
Cellulose loose-fill	3.7 to 4 (when poured), 3.1 to 4 (when blown)
Vermiculite loose-fill	2 to 2.6
Perlite loose-fill	2 to 2.7
Polystyrene rigid	4 to 5.4
Polyurethane rigid	6.7 to 8
Polyisocyanurate rigid	8

Regardless of whether you install your own insulation or sub the work out, you have to know enough about the various products to satisfy your customers. You might go for months or even years without having a customer ask for a detailed comparison of the types of insulation available. However, if this question comes up during a meeting between a prospective customer and yourself, you could lose a lot of credibility if you are unprepared to answer it. Let me give you a quick example of what I mean.

Are you ready to field questions about insulation? How much do you know about it? Let's say I am a customer and I have a question.

TABLE 16.2 Ratings for fiberglass batt insulation.

Thickness (inches)	R-value
3½	11
3⅜	13
6½	19
7	22
9	30
13	38

TABLE 16.3 Ratings for rock wool batt insulation.

Thickness (inches)	R-value
3½	11
3⅝	13
6½	19
7	22

My first question pertains to the attic insulation you will be installing. Will it be in batts, or will it be loose-fill insulation. You tell me that it will be loose-fill. Then I ask if the insulation will be blown into place or installed by hand. You stumble over words, trying to think while you talk. It appears to me that you have not considered the method of installation. This concerns me. Since you don't know how the insulation will be installed, how do you know how much to charge me? Have you just pulled your prices out of thin air?

My next question requires you to describe the type of loose-fill material that will be used. Will it be a glass-fiber product, mineral wool, or cellulose? You are again at a loss for words, and this makes my impression of your professionalism dwindle. It is becoming obvious to me that you don't know much about the insulating work that will be done for me.

Even though I'm guessing that you are not up to speed, I ask you to explain the pros and cons of mineral wool insulation. I go on to ask for a detailed evaluation of how glass-fiber insulation stacks up against cellulose. Your lack of skilled responses is really starting to worry me. Should I eliminate you from consideration simply because

TABLE 16.4 Ratings for loose fill insulation.

Type	R-value per inch of insulation
Cellulose	2.8 to 3.7
Fiberglass	2.2 to 4.0
Perlite	2.8
Rock wool	3.1
Vermiculite	2.2

fastfacts

Don't lose customers or good references from a lack of product knowledge. Get to know the product that you are selling.

you are not fluent in insulation details? I probably shouldn't, but I might. You could be the best general contractor in the area, but if you impress me as someone who doesn't take technical issues seriously, I may not trust you to see my job through to a successful completion. By not knowing the answers to my questions pertaining to insulation, you could lose the entire job. The fact that you rely on professional insulation contractors to advise you on what types of insulation to use could be lost on me. If these professionals were present to answer my questions, there wouldn't be any problem. But, since you are alone and unable to give solid responses to my questions, I'm definitely going to lose some confidence in you. This is something no contractor can afford to have happen. To avoid getting boxed into a corner, you need enough knowledge of insulation to carry on a competent conversation. Let's examine the various types available and what their prime uses are.

Glass-Fiber Insulation

Glass-fiber insulation is by far the most widely used type of residential insulation. It is installed in crawl spaces, exterior walls, and attics. The insulation is available in various forms. You might buy loose-fill material, faced batts, unfaced batts, and so forth. R-values vary based on the thickness of the insulation. As well known and popular as glass-fiber insulation is, it is also one of the more difficult types to work with if you have sensitive skin. Glass-fiber insulation makes a lot of people itch. This condition is usually worst in hot weather, but it can occur at any time of the year. Aside from the irritating nature of glass-fiber insulation, the remainder of its features and benefits are basically good.

Batts

Batts and blankets of glass-fiber insulation are used in attics and exterior wall cavities. Gaining access to an attic to install batts and blankets is usually easy, but to insulate a wall with this type of material, the entire wall with have to be opened up. In some remodeling jobs this isn't a problem, since all of interior wall coverings may be removed from exterior walls. If destroying wall coverings will be a problem, you can still use glass-fiber insulation.

Loose-Fill

Glass-fiber insulation is available in a loose-fill form. This type of insulation can be spread around an attic by hand, or it can be blown into attics and exterior walls. Unlike batts, where entire walls have to be opened for installation, loose-fill material can be blown into wall cavities through small holes. The holes are much easier and less expensive to repair than a complete rip-out of the wall coverings.

Rigid Boards

You might not expect to find glass-fiber insulation in the form of rigid boards. These boards are used to add insulating value to exterior walls that are being constructed. The rigid boards are also used to insulate basement walls, both inside and out. They can even be used to help insulate vaulted ceilings where there is no attic. In terms of rigid insulation, glass-fiber doesn't stack up well against its competitors, polystyrene and urethane.

Mineral Wool

Mineral wool is in many ways similar to glass-fiber insulation. This insulation is available in batts, blankets, and loose-fill. Its R-Value per inch of insulation is the same as that of glass-fiber insulation. Mineral wool is by no means, a poor insulator, but if you don't like to itch and you want a slightly higher R-value, cellulose might be worth considering.

TABLE 16.5   Sizes of insulation boards.

Widths	Lengths	Thickness
4 feet	4 feet to 16 feet	½"
		⅝"
		¾"
		1"

Cellulose

Cellulose insulation is limited in its use. Since the product is available only as a loose-fill insulator, it is not practical to install cellulose in the stud bays of new construction. If you will be blowing insulation into existing walls, cellulose is a worthy contender. It's R-value rating is slightly better than mineral wool and glass-fiber insulation. There is both good and bad to assess with cellulose.

If cellulose insulation gets wet, it loses much of its insulating quality. Untreated cellulose is also a considerable fire hazard. When you plan to install cellulose insulation, make sure that it has been treated to be fire-resistant. Old paper is the prime ingredient in cellulose insulation. Knowing the properties of paper, you can imagine how cellulose insulation performs when it is subjected to extended moisture, insects, rodents, and so forth. It doesn't fare well.

The strong points to cellulose are that it is affordable and will not normally cause irritation for installers. There is also the fact that there is little to worry about in terms of odor emissions or health threats.

Polystyrene

Polystyrene is used in the construction of rigid insulation boards. The insulating quality of polystyrene is very good. The downside to this insulator is its cost and the fact that it can be flammable. The R-value for polystyrene is the same as that of glass-fiber insulation and mineral wool. All these insulation materials share a rating of R-3.5 for every inch of insulation installed.

Urethane

Urethane is one of the most effective insulators available, but it is illegal to use in some locations. If you consider that most insulation materials have a value of around R-3.5, urethane has a rating of R-5.5. Unfortunately, urethane is also known to produce cyanide gas if it burns. This, of course, is a deadly gas. Due to the potential risk of creating a poisonous gas, urethane insulation has been banned in a number of locations.

Urethane is available in the form of rigid boards and as a foam. The foam version was extremely popular for old homes made of brick and block. If allowed by local codes, urethane is far and away the most effective insulation material you can use in terms of R-values. But, you aren't likely to fill the stud bays of a new addition with

fastfacts

Rigid foam has become a very popular part of new construction. Check out the new products and see what is out there. Using rigid foam with normal wall-cavity insulation can give your buildings very impressive energy-efficiency numbers.

foam. You have to plot your work in accordance with your personal circumstances.

R-VALUES

R-values are a unit of measurement intended to establish the resistance of a certain material. The higher the resistance level is, the better the insulating quality of the material. For example, insulation with a rating of R-19 is not as good as an R-30 insulation. Most homeowners are familiar with R-values, so you shouldn't be forced to educate many people on what they are or how they work. However, you might get some questions pertaining to existing building materials and their R-values.

Do you know what the average R-value of a single-glazed window is? I know that windows are normally rated with U-values, where the lower the rating the better the window. The R-value of an uninsulated window is about R-1. A double-glazed window should have a rating in the neighborhood of R-2. What would you guess the R-value of an average older door to be? If you guessed R-1, you should be in the ballpark. Storm doors can raise the rating to R-2.

When customers are talking with you about new insulation, they may want an idea of what their existing building components are doing for them in terms of R-value. A typical exterior wall in a wood frame house that is covered with wood siding will carry a rating of around R-5. If an insulating sheathing has been installed beneath the siding, this rating could go up to R-7. If a home has an 8-inch brick wall, the R-value will probably be around R-4.

Ceilings that are made of drywall normally carry an R-value of 4. This rating is subject to the attic conditions if there is an attic over the drywall ceiling. It is not unreasonable to assume an R-value of 8

TABLE 16.6 R-values for common building materials.

Material	R-value per inch or as specified
Concrete	0.11
Mortar	0.20
Brick	0.20
Concrete block	1.11 for 8"
Softwood	1.25
Hardwood	0.090
Plywood	1.25
Hardboard	0.75
Glass	0.88 for single thickness
Double-pane insulated glass	1.72–⅜ w¼" air space
Air space (vertical)	1.35–¾"
Gypsum lath and plaster	0.40–⅞"
Dry wall	0.35–½"
Asphalt shingles	0.45
Wood shingles	0.95
Slate	0.05
Carpet	2.08
Vinyl flooring	0.05

in some circumstances. You can't arrive at an accurate R-value unless you know what all the existing materials are and the conditions surrounding them.

Floors made of wood might carry a rating of R-4. If carpeting is installed over the floor, this rating might hit R-6. Most houses, even old ones, will have some insulation in them. Attics and crawl spaces are the most likely areas to find this insulation. Since access is better for an attic or floor than it is for an exterior wall, these locations usually get top billing when it comes to doing an energy upgrade. To evaluate the R-value of insulation, you must have some means of measurement. This is typically done by measuring the depth of the insulation and converting the depth to an R-value. While this is more common in remodeling jobs, it applies to new construction. To do this conversion, you need some numbers to plug in. Here they are. (All R-values are based on 1 inch of insulation.)

- Glass-fiber   R-3.5
- Mineral wool   R-3.5
- Cellulose   R-3.6
- Vermiculite   R-2.2
- Perlite   R-2.4
- Polystyrene   R-3.5
- Urethane foam   R-5.5

VAPOR BARRIERS

Vapor barriers are needed when installing insulation. Without them moisture can build up in wall cavities and cause wood products to rot. Mildew is another potential side effect. There are several ways to create a vapor barrier. Many manufacturers offer both faced and unfaced batts of insulation. The facing on a batt of insulation serves as a vapor barrier. Insulation contractors frequently install unfaced insulation and then cover it with plastic. This also creates a vapor barrier.

Condensation can be a big problem in some houses. When condensation forms, moisture is present. This moisture can cause a house to deteriorate before its time. If a proper vapor barrier is not installed, condensation can rot wood structural members and reduce the efficiency of insulation to half its normal R-value rating. This damage generally occurs over a number of years and is not normally found until significant structural damage has been done

Towards the Heated Space

Vapor barriers should be installed toward the heated space of a home. If you are using faced insulation, the facing should be visible from the living space of the house prior to being covered with drywall. Installing insulation backwards, and I've seen jobs where this has been done, will result in some serious moisture problems.

I remember an article in a newspaper showing how a nearly new house had rotted because the vapor barriers on the outside walls had been installed backwards. Instead of repelling moisture back into the house, these backwards barriers trapped water and caused it to saturate the insulation. The result was ruined insulation, rotted wood, and a very, very unhappy customer.

fastfacts

It is common for customers to want tight, energy-efficient homes. In my opinion, this can be carried to an extreme. A house that is too tight might not be healthy. I am not an expert in this field, but I am pretty sure that a routine air exchange is needed to maintain suitable living conditions.

From personal experience, I've found most mistakes with vapor barriers to be made in crawl spaces. This is the one location where I have personally encountered insulation installed with the vapor barrier upside down. Whether these installations were done by professional contractors or homeowners I don't know, but I do know that the jobs had been done incorrectly.

VENTILATION

Ventilation is needed in a home, and installing insulation and vapor barriers can reduce ventilation to a point where air inside the home may not be healthy. The current construction field has undergone numerous changes over the years. Some have proved to be good, and others have not enjoyed such enviable track records. One mistake learned during this time is that it is possible to make a house too tight. If air is not allowed to come and go in a house, big problems can crop up. As a contractor, you should be aware of these potential problems.

Are you aware that carpeting and furniture can emit dangerous substances? Do you know what radon is and how it affects people? How much do you know about the vapors and fumes that may accumulate during an average day's cooking in a kitchen? Going to an extreme, how long can a person breathe stale air before the oxygen level is depleted? All these questions pertain to what we're talking about. If a house is sealed up too tightly, any of the issues we have just touched on can grow in magnitude.

How many houses get wrapped before they get sided? A lot. Would you assume that most newer houses are filled with insulation and secured with plastic vapor barriers? I would. Are today's windows and doors tighter than the ones in your grandparent's home when you were a toddler? They certainly are. With all the fuss to create a more efficient home, some builders have created monsters. The houses they built are too tight. Don't allow yourself to fall into this same trap. Give your customers a good job, but don't seal them in so tightly that they will suffer from the potential consequences.

17

ENGINEERED WOOD PRODUCTS

Solid wood products, such as framing lumber and wood paneling, have limitations with respect to dimensional stability, uniformity and structural integrity–nature rarely if ever makes
two identical things. So called *engineered wood products* are just
that, products manufactured of wood that are uniform, dimensionally stable, and structurally predictable by utilizing the best
properties of selected wood and wood components in the production of these products.

These engineered wood products also utilize what might otherwise be considered scrap or waste from logging and milling operations. Products such as waferboard, particleboard, and the more
popular oriented strand board (OSB) consist of by-products such as
strips or chips of wood.

Everyone is familiar with plywood, although maybe not all of
the variations and grades of this versatile product along with the
wide choice of veneers for uses ranging from concrete forms to
Board Room paneling. Laminated beams referred to as Glu-Lams,
I-joists so named because when viewed from their ends, they look
like a capital "I", medium density particleboard (MDF) more familiar under the Masonite™ name are all part of the engineered
wood product family.

PLYWOOD

With multiple veneer grades, surface treatments, and textures, plywood can have widely different functions and applications.

Veneers

Let's start with veneer grades, the first step in selecting the right panel:

- Grade A: smooth, printable, not more than 18 neatly made repairs; may be used in natural finish in some applications; subcategories: A-A: both faces are A quality with D-grade interior; A-A Exterior: C-grade inner ply bonded together with exterior glue; A-B: Grade A and B faces with D-grade inner plies, used as substitute for A-A when only one face will be exposed.; A-B, A-C Exterior, A-D Interior are other grades.

- Grade B: solid surface with tight knots to 1 inch (2.54cm) across grain wood or synthetic repairs, and some minor splits allowed.; subcategories: B-B: both faces B grade with D-grade inner plies; B-B and B-C Exterior: inner plies of C grade bonded with exterior glue; B-B Plyform™ for concrete formwork, mill-oiled unless specified otherwise.

- Grade C: veneer with splits allowed up to ⅛ inch (3.175mm) wide and knot holes or other open defects limited to ¼ x ½ inch (6.35mm x 12.5mm); repairs and some broken grain permitted; subcategories: C-C Plugged Exterior: faces C grade with C inner plies bonded with exterior glue, used on exterior decks or balconies, boxcars, and truck floors; C-D Plugged: C-Plugged-grade face, D-grade back, bonded with either interior or exterior glue, used for cable reels, walkways.

- Grade D: knots and knotholes to 2½ inches (6.35mm) in width across grain and ½ inch (1.25mm) or larger, within specified limits, stitching permitted; subcategories: CDX: exterior grade panel designed for exposure to the weather or to moisture for its entire life.

There are also several exposure durability classifications:

- Exposure 1: Panels that are fully waterproofed, designed for long-term exposure where high moisture conditions are anticipated.

- Exposure 2: Interior panels with intermediate glue (moisture-resistant but not waterproof) intended for construction applications

fastfacts

Exterior grade Exposure 1 panels have a fully waterproof bond but are designed to be used where construction delays may delay covering the panel, leaving it exposed to the weather for possibly six months.

Exterior panels have a fully waterproof bond and are meant for long term exposure to the weather.

where limited protection from moisture may be anticipated; Exposure 2 Exterior panels with designation PS-1 have an adhesive that is 100% waterproof

When "repairs" are mentioned in plywood specifications, they refer to patches or plugs inserted in the veneer to replace a defect. Boat patches, shaped like a small rowboat, are most common, but there are others, including router patches, with parallel sides and rounded corners; shim patches, long, narrow wood inserts; plug or dog bone patches as well as circular patches are used along with synthetic resin to fill in some of the defects in the veneers (Figure 17.1).

COM-PLY™ is a composite panel manufactured by bonding layers of wood fibers between wood veneer. These panels are made in three- or five-layer arrangements; the three-layer panel has a wood fiber core and veneer face and back; the five-layer panel has a veneer crossband in the center.

Sheathing and Siding

Sheathing can be purchased with span ratings from 12/0 to 16/0, 24/0, 24/16, 32/16, 40/20, and 48/24. It comes in thicknesses from ⅝ inch to ¾ inch (15.87mm to 19.04mm) and with three exposure ratings, Exterior, Exposure 1, and Exposure 2. Note: span ratings refer to maximum center-to-center length of supports. The left-hand number represents the maximum recommended center-to-center spacing of supports in inches; the right-hand number is the maximum center-to-center support in inches when the panel is used for subflooring with long dimension (say the 8-foot dimension on a 4 x 8 panel) across support. A left-hand number of 24 means

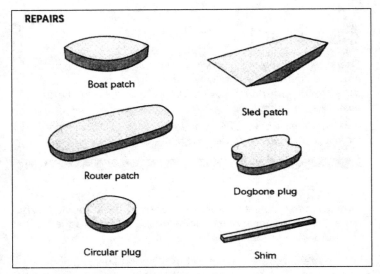

REPAIRS

Boat patch

Sled patch

Router patch

Dogbone plug

Circular plug

Shim

FIGURE 17.1 Typical plywood "repairs".

that the panel can be applied to studs spaced every 24″ (60.96 cm), referred to as 24 inches (60.96) on center.

Sheathing is the structural covering on the outside surface of the structure, providing support for facing materials such as siding , brick, or stone veneer and roofing materials.

Plywood used as a finished siding material is available in a number of textures:

- Brushed or relief grain: available in Douglas fir or cedar veneer; thickness availability: ¹¹⁄₃₂ (8.73mm), ⅜ (9.5mm), ¹⁵⁄₃₂ (11.9 mm), ½ (1.25cm) and ⅝ inch(15.87mm)

- Channel groove: grooves ¹⁄₁₆″ (1.58mm) or ⅜″ (9.5mm) are cut into the face of the plywood, either 4 or 8 inches (10.16 cm or 20.32 cm) on center; available in Douglas fir, cedar, southern pine, and some other species; thickness: ¹¹⁄₃₂″ (8.73mm), ⅜″ (9.5mm), ¹⁵⁄₃₂″ (11.9mm), ½″ (1.25cm)

- Medium-density overlay: with or without grooving, smooth or textured surface., available in board and batten or T-111 configuration; thickness: ¹¹⁄₃₂″ (8.73mm), ⅜″ (9.5mm), ¹⁵⁄₃₂″ (11.9mm), ½″ (1.25cm), ¹⁹⁄₃₂″ (15.08mm), or ⅝″ (15.87mm) butt or lap joint

- Reverse board and batten: deep grooves, ¼ x 1 (6.35mm x 25.4mm) to 1 1/2 "(38.1mm) wide, spaced 8" (20.32cm) or 12 "(30.48cm) on center, available in several wood species including Douglas fir, southern Pine, and cedar; thickness: ¹⁹⁄₃₂" (15.08mm), ⅝" (15.87mm)

- Lap siding: square or beveled edge, rough sawn or smooth face, cut into lap siding dimension; thickness: ¹¹⁄₃₂" (8.73mm), ⅜" (9.5mm), ¹⁵⁄₃₂" (11.9mm), ½" (1.25cm), ¹⁹⁄₃₂" (15.08mm), and ⅝" (15.87mm)

Fire-Retardant Plywood

Fire-retardant panels are pressure-impregnated with chemicals mixed in water per American Wood Preservers Association standards. The degree of fire retardation should be verified by contacting the treatment service.

Marine Grade Plywood

Marine plywood is manufactured of Douglas fir and western larch with faces of B grade or better. The panels are sanded on both sides, and no core gap exceeding ⅛ inch (3.175mm) permissible. The adhesive in marine plywood is a fully waterproof material. Grades A-A, A-B, and B-B are available, and panels may be overlaid to high or medium density (see below).

High Density Overlay (HDO) Panels

These panels are finished with a resin-impregnated overlay to create a hard, very smooth surface that can be painted with minor additional preparation. The panels can be used for such varied purposes as highway signs and cabinets.

Medium Density Overlay (MDO) Panels

These plywood panels are finished with an opaque resin-treated material that provides an excellent painting surface and can be used as siding, for built-ins, signs, and displays, and even some types of furniture. MDO panels are available with a smooth face, V grooved, Texture 1-11, or board and batten grooving.

fastfacts

*Do not butt sheathing panels together; leave at least 1/8 inch
(3.175mm) between joints.*

HARDBOARD (COMPRESSED FIBERBOARD)

A board manufactured from interfelted lignocellulosis fibers consolidated under heat and pressure to form a dense material.

- Available thickness: Typically ½ inch (12.7 mm) to 2 inches (50.8 mm)
- Density: Typically 10 to 25 pounds/cubic foot (160 to 400 kg/cubic meter)
- Uses: Wall sheathing, roof insulation, and sound insulation

ORIENTED STRAND BOARD (OSB)

This material originated from waferboard and is constructed of rectangular strands of softwood or hardwood ½ inch (12.7 mm) by 3 inches (76.2 mm) to 4 feet 6 inches (1.37 m) in length laid up in mats to form a panel arranged in layers at right angles to each other. This cross orientation makes the panels strong and stiff in both directions. OSB is bonded together with fully waterproof adhesives, and most panels are treated with an edge sealant to retard moisture penetration.

- Available thickness: Typically ¼ inch (6.4 mm) to 1⅛ inches (28.6 mm)
- Density: 36 to 44 pounds/cubic foot (577 to 705 kg/cubic meter)
- Uses: In structural applications in the same way as plywood; phenolic overlay OSB is used for siding

WAFERBOARD

Similar to OSB, except that it is composed of large flakes of wood bonded together and generally made of low-density hardwoods, such as aspen. Once used a great deal as sheathing, it has largely been replaced by OSB.

LAMINATED VENEER LUMBER (LVL)

This board is primarily a structural member made of veneer laid up in one grain direction and made in billets 27 inches (68.6 cc) to 50 inches (127 cm) wide and 1½ inches (38.1 mm) to 1¾ inches (44.5 mm) thick. It is produced under pressure to cure the adhesives, mostly phenolic glues. This material is non-destructively tested to ensure consistent strength. Some manufacturers use LVLs as flanges for truss joists. Typical uses are rafters, headers, beams, joists, studs, and columns.

PARALLEL STRAND LUMBER (PSL)

This product is made of oriented strands of waste softwood veneer. The ½ inch (12.7 mm) wide by 37 inches (94 cm) long strands are oriented and laid up in a mat, which is processed through a microwave heating system into billets 11 inches (279 mm) x 18 inches (457 mm) or 11 inches (279 mm) x 14 inches (356 mm). These billets are sawn into lengths and thicknesses as required. PSL members are used where high-strength lumber is required, and some builders use them as components in wood joist construction.

GLULAMS

Glulams, engineered ridge, roof, and floor beams, are pound for pound stronger than steel and have greater structural strength than comparably sized dimensional lumber. Able to span greater distances without intermediate support, glulams are often used as architecturally exposed framing members and are available in four appearance groups: framing, industrial, architectural, and premium.

Glulams are the only engineered wood product that can be easily cambered to ensure that the beam will not sag or be subject to excessive deflection under gravity loads. Glulam roof beams are often cambered for 1.5 calculated dead-load deflection. For floor beams the camber recommended is 1.0 times the calculated dead-load deflection.

Glulams can be used for stair stringers when a long stringer is required, and custom curved glulams are also available. Glulams as columns can replace spliced dimensional lumber when long lengths are required, and, by selecting a specific wood species, can be left exposed as an architectural treatment

Sealers must be applied to glulams, either in the mill or at the job site, to prevent excessive moisture penetration. Surface sealants also assist in controlling checking and grain raising. Checking can occur in the various layers that make up the glulam, but elastomeric fillers can be used to repair and hide any small checks.

Properly designed connections are key to the performance of a glulam for not only holding power but also to compensate for any minor swelling of the wood.

Glulams are available in a number of widths and depth.

Douglas fir glulam equivalents are shown in Figure 17.2. Figures 17.3 through 17.10 show suggested placement and framing details incorporating glulams.

I-JOISTS

An I-joist is a wood and wood composite joist (floor or roof) that looks like an "I." The top and bottom chords, also known as flanges, are made of lumber or laminated veneer lumber (LVL). The web (center section) can be made of plywood or OSB. I-joist standard sizes, depth, and flange widths are shown in Figure 17.11.

I-joists should not be stored directly on the ground, and they need to be protected from the weather so care must be taken when they are "picked" during placement. APA-The Engineered Wood Products Association recommends picking in bundles as shipped by the manufacturer; orient the bundles so that the webs of the I-joists are vertical and pick the bundle at the fifth points using a spreader bar if necessary. Don't twist or apply any loads to the I-joists when they are horizontal

I-joists are not stable until completely installed and will not support or carry a load until they are fully braced and sheathed. To avoid accidents, brace and nail each I-joist as it is installed using hangers, rim board, or blocking.

Figures 17.12 through 17.15 illustrate some framing details utilizing I-joists.

Equivalent Douglas Fir Glulam Sections as Substitutes for Sawn Lumber

EQUIVALENT DOUGLAS FIR GLULAM SECTIONS AS SUBSTITUTES FOR DOUGLAS-FIR SAWN LUMBER

Sawn Section (Nominal)	Floor Beams (100%) No. 1	Equivalent Glulam Sections Roof Beams (115%) No. 1	Roof Beams (125%) No. 1
3 × 10	3-1/8 × 9	3-1/8 × 9	3-1/8 × 9
3 × 12	3-1/8 × 10-1/2	3-1/8 × 9	3-1/8 × 10-1/2
3 × 14	3-1/8 × 12	3-1/8 × 10-1/2	3-1/8 × 10-1/2
4 × 10	3-1/8 × 10-1/2	3-1/8 × 10-1/2	3-1/8 × 10-1/2
4 × 12	3-1/8 × 12	3-1/8 × 10-1/2	3-1/8 × 10-1/2
4 × 14	3-1/8 × 13-1/2	3-1/8 × 12	3-1/8 × 12
6 × 10	5-1/8 × 10-1/2	5-1/8 × 10-1/2	5-1/8 × 10-1/2
6 × 12	5-1/8 × 12	5-1/8 × 12	5-1/8 × 12
6 × 14	5-1/8 × 13-1/2	5-1/8 × 13-1/2	5-1/8 × 13-1/2
6 × 16	5-1/8 × 15	5-1/8 × 13-1/2	5-1/8 × 15
2 – 2 × 8	3-1/8 × 7-1/2	3-1/8 × 7-1/2	3-1/8 × 7-1/2
2 – 2 × 10	3-1/8 × 9	3-1/8 × 9	3-1/8 × 9
2 – 2 × 12	3-1/8 × 10-1/2	3-1/8 × 10-1/2	3-1/8 × 10-1/2
2 – 2 × 14	3-1/8 × 12	3-1/8 × 10-1/2	3-1/8 × 10-1/2
3 – 2 × 8	3-1/8 × 9	3-1/8 × 9	3-1/8 × 9
3 – 2 × 10	3-1/8 × 10-1/2	3-1/8 × 10-1/2	3-1/8 × 10-1/2
3 – 2 × 12	3-1/8 × 13-1/2	3-1/8 × 12	3-1/8 × 12
3 – 2 × 14	3-1/8 × 13-1/2	3-1/8 × 13-1/2	3-1/8 × 13-1/2

Notes:
(1) Span = uniformly loaded simply supported beam with a span ranging from 8 ft up to 20 ft.
(2) For roof beams, maximum deflection = L/180 under total load. Deflection under live load must be verified when live load/total load > 3/4.
(3) For floor beams, maximum deflection = L/360 under live load, based on live load/total load = 0.8. Where additional stiffness is desired or for other live load/total load ratios, design for deflection must be modified per requirements.
(4) Service condition = dry.
(5) Beam weights for solid-sawn and glulam members are assumed to be the same.
(6) Design properties at normal load duration and dry-use service conditions – No. 1: $F_b = C_F \times 1{,}000$ psi, $F_c = 95$ psi, $E = 1.7 \times 10^6$ psi, where C_F = size factor per 1991 NDS. Glulam: $F_b = C_v \times 2{,}400$ psi, $F_v = 190$ psi, $E_x = 1.8 \times 10^6$ psi, where C_v = volume factor per 1991 NDS. Repetitive member factor is assumed to be 1.15 for the 3-member built-up lumber beams and 1.0 for 2-member built-up lumber beams and glulam beams.

(By permission from APA—The Engineered Wood Association, Tacoma, Washington.)

FIGURE 17.2 Douglas Fir–Glulam equivalents.

(text continues on page 291)

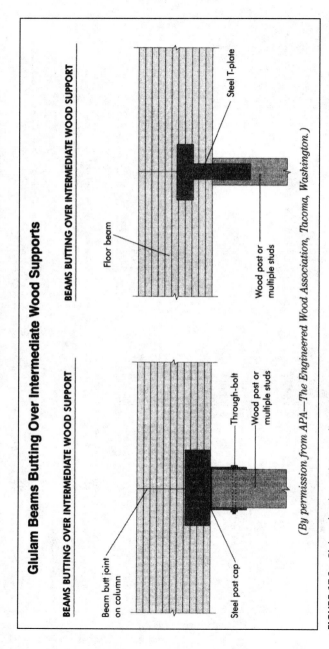

FIGURE 17.3   Glulams butting at intermediate supports.

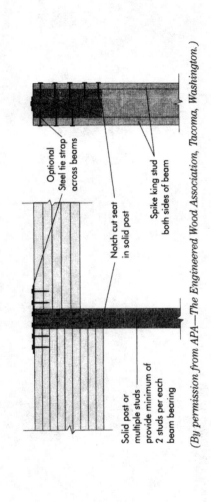

Beam Size Changes Over Intermediate Supports

BEAM SIZE CHANGE OVER INTERMEDIATE SUPPORT

Optional
Steel tie strap
across beams

Notch cut seat
in solid post

Spike king stud
both sides of beam

Solid post or
multiple studs — provide minimum of
2 studs per each
beam bearing

(By permission from APA—The Engineered Wood Association, Tacoma, Washington.)

FIGURE 17.4 Glulams changing sizes over intermediate supports.

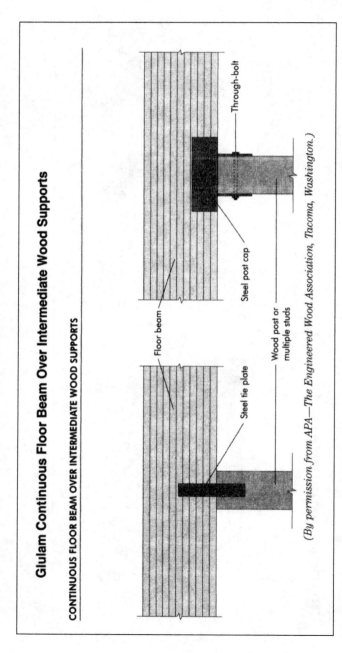

FIGURE 17.5 Glulams continuous over intermediate supports.

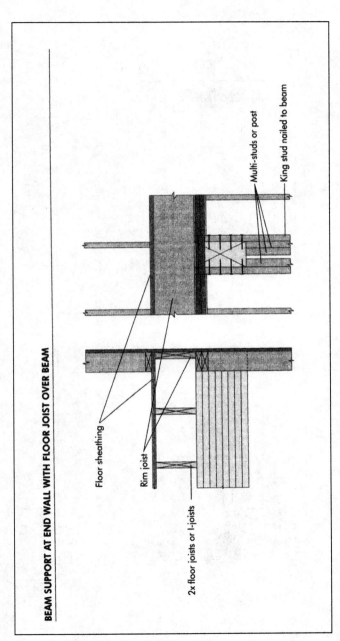

BEAM SUPPORT AT END WALL WITH FLOOR JOIST OVER BEAM

Multi-studs or post

King stud nailed to beam

Floor sheathing

Rim joist

2x floor joists or I-joists

FIGURE 17.6 Glulams bearings at end walls with steel tie cap plates.

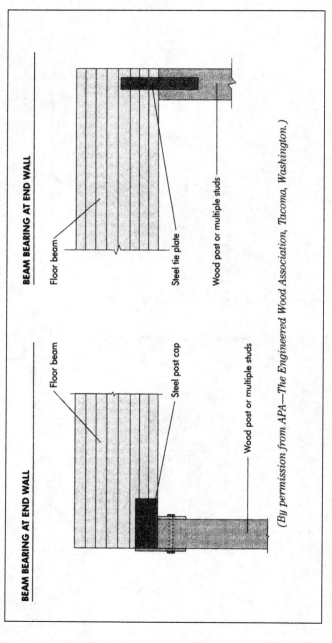

FIGURE 17.7 Glulam bearing on end walls.

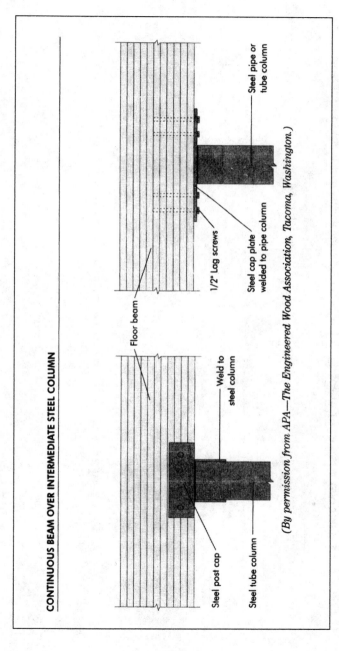

CONTINUOUS BEAM OVER INTERMEDIATE STEEL COLUMN

Floor beam

1/2" Lag screws

Steel cap plate welded to pipe column

Steel pipe or tube column

Weld to steel column

Steel post cap

Steel tube column

(By permission from APA—The Engineered Wood Association, Tacoma, Washington.)

FIGURE 17.8 Continuous glulam beam over intermediate steel column.

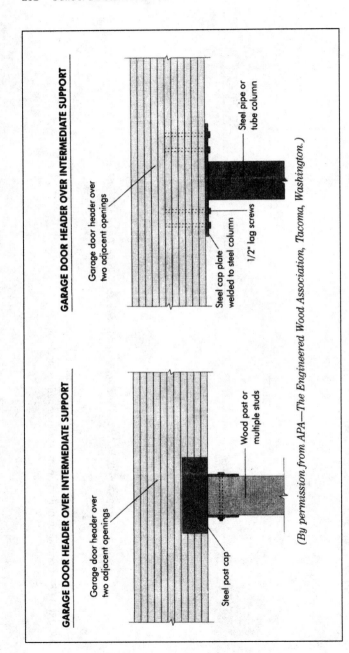

FIGURE 17.9 Glulams as garage door headers.

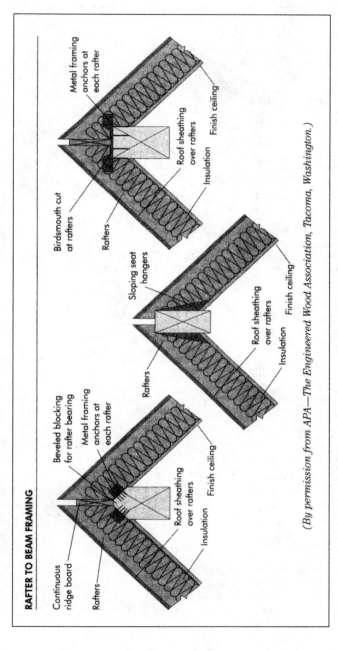

RAFTER TO BEAM FRAMING

Continuous ridge board

Rafters

Roof sheathing over rafters

Insulation

Finish ceiling

Beveled blocking for rafter bearing

Metal framing anchors at each rafter

Rafters

Sloping seat hangers

Roof sheathing over rafters

Insulation

Finish ceiling

Birdsmouth cut at rafters

Rafters

Metal framing anchors at each rafter

Roof sheathing over rafters

Insulation

Finish ceiling

(By permission from APA—The Engineered Wood Association, Tacoma, Washington.)

FIGURE 17.10 Rafter to beam framing.

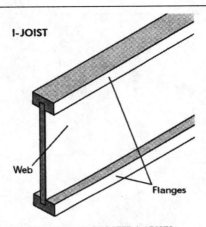

DIMENSIONS FOR APA PERFORMANCE RATED I-JOISTS

APA PRI

Joist Series	Joist Designation		Nominal Depth	Net Depth	Flange Width
I x 10	C4	(PRI-15)	10"	9-1/2"	1-1/2"
I x 10	C6	(PRI-25)	10"	9-1/2"	1-3/4"
I x 12	C10	(PRI-15)	12"	11-7/8"	1-1/2"
I x 12	C12	(PRI-25)	12"	11-7/8"	1-3/4"
I x 14	C14	(PRI-25)	14"	14"	1-3/4"
I x 14	C16	(PRI-35)	14"	14"	2-5/16"
I x 16	C18	(PRI-25)	16"	16"	1-3/4"
I x 16	C20	(PRI-35)	16"	16"	2-5/16"
I x 10	S2	(PRI-30)	10"	9-1/2"	2-1/2"
I x 10	S4	(PRI-32)	10"	9-1/2"	2-1/2"
I x 12	S6	(PRI-30)	12"	11-7/8"	2-1/2"
I x 12	S8	(PRI-32)	12"	11-7/8"	2-1/2"
I x 12	S10	(PRI-42)	12"	11-7/8"	3-1/2"
I x 14	S12	(PRI-32)	14"	14"	2-1/2"
I x 14	S14	(PRI-42)	14"	14"	3-1/2"
I x 16	S16	(PRI-32)	16"	16"	2-1/2"
I x 16	S18	(PRI-42)	16"	16"	3-1/2"

Notes:
1. Tolerances permitted at time of manufacture:
   - flange width = +/− 1/32 inch
   - I-joist depth = + 0 inches, − 1/8 inch

(By permission from APA—The Engineered Wood Association, Tacoma, Washington.)

FIGURE 17.11 Standard sizes for I-joists.

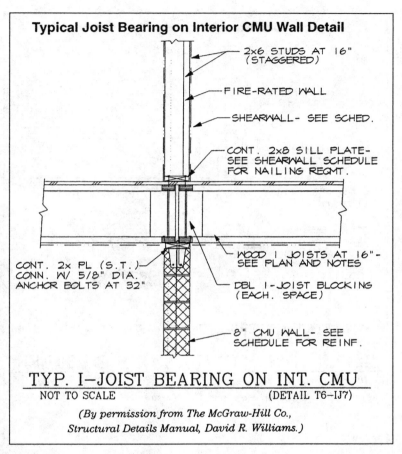

Typical Joist Bearing on Interior CMU Wall Detail

2x6 STUDS AT 16"
(STAGGERED)

FIRE-RATED WALL

SHEARWALL- SEE SCHED.

CONT. 2x8 SILL PLATE-
SEE SHEARWALL SCHEDULE
FOR NAILING REQMT.

CONT. 2x PL (S.T.)
CONN. W/ 5/8" DIA.
ANCHOR BOLTS AT 32"

WOOD I JOISTS AT 16"-
SEE PLAN AND NOTES

DBL I-JOIST BLOCKING
(EACH. SPACE)

8" CMU WALL- SEE
SCHEDULE FOR REINF.

TYP. I–JOIST BEARING ON INT. CMU
NOT TO SCALE   (DETAIL T6–IJ7)

*(By permission from The McGraw-Hill Co.,
Structural Details Manual, David R. Williams.)*

FIGURE 17.12 I-joist bearing on interior CMU wall.

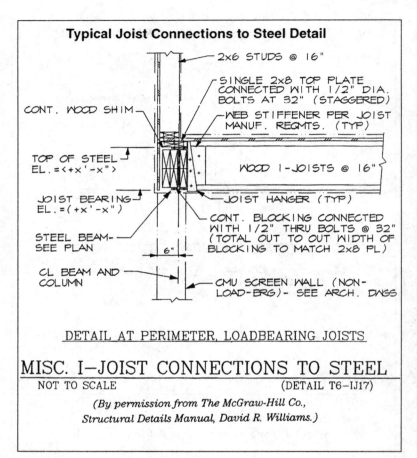

Typical Joist Connections to Steel Detail

2x6 STUDS @ 16"

SINGLE 2x8 TOP PLATE CONNECTED WITH 1/2" DIA. BOLTS AT 32" (STAGGERED)

CONT. WOOD SHIM

WEB STIFFENER PER JOIST MANUF. REQMTS. (TYP)

TOP OF STEEL EL.=<+x'-x">

WOOD I-JOISTS @ 16"

JOIST BEARING EL.=(+x'-x")

JOIST HANGER (TYP)

STEEL BEAM— SEE PLAN

6"

CONT. BLOCKING CONNECTED WITH 1/2" THRU BOLTS @ 32" (TOTAL OUT TO OUT WIDTH OF BLOCKING TO MATCH 2x8 PL)

CL BEAM AND COLUMN

CMU SCREEN WALL (NON-LOAD-BRG)- SEE ARCH. DWGS

DETAIL AT PERIMETER, LOADBEARING JOISTS

MISC. I—JOIST CONNECTIONS TO STEEL

NOT TO SCALE   (DETAIL T6-IJ17)

(By permission from The McGraw-Hill Co., Structural Details Manual, David R. Williams.)

FIGURE 17.13 I-joist connecting to steel.

Typical Joist Connections to Steel—at Roof and Floor Level

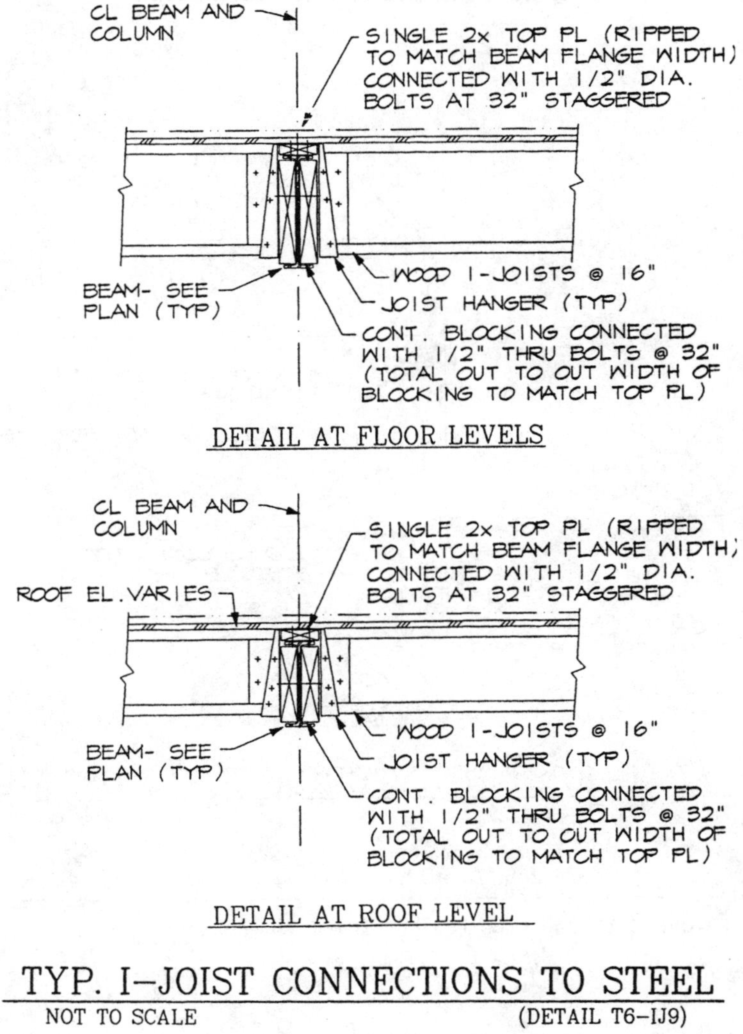

CL BEAM AND COLUMN

SINGLE 2x TOP PL (RIPPED TO MATCH BEAM FLANGE WIDTH) CONNECTED WITH 1/2" DIA. BOLTS AT 32" STAGGERED

BEAM- SEE PLAN (TYP)

WOOD I-JOISTS @ 16"

JOIST HANGER (TYP)

CONT. BLOCKING CONNECTED WITH 1/2" THRU BOLTS @ 32" (TOTAL OUT TO OUT WIDTH OF BLOCKING TO MATCH TOP PL)

DETAIL AT FLOOR LEVELS

CL BEAM AND COLUMN

ROOF EL. VARIES

SINGLE 2x TOP PL (RIPPED TO MATCH BEAM FLANGE WIDTH) CONNECTED WITH 1/2" DIA. BOLTS AT 32" STAGGERED

BEAM- SEE PLAN (TYP)

WOOD I-JOISTS @ 16"

JOIST HANGER (TYP)

CONT. BLOCKING CONNECTED WITH 1/2" THRU BOLTS @ 32" (TOTAL OUT TO OUT WIDTH OF BLOCKING TO MATCH TOP PL)

DETAIL AT ROOF LEVEL

TYP. I–JOIST CONNECTIONS TO STEEL

NOT TO SCALE   (DETAIL T6–IJ9)

(By permission from The McGraw-Hill Co., Structural Details Manual, David R. Williams.)

FIGURE 17.14 I-joist connections to roof/floor steel.

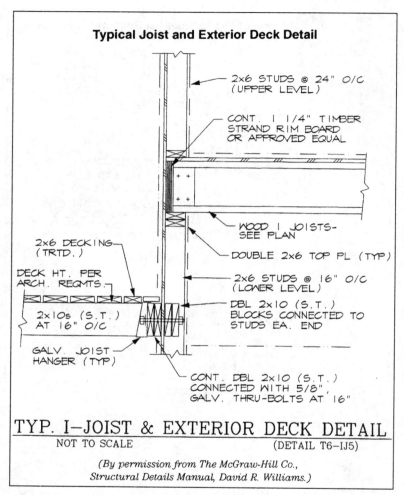

Typical Joist and Exterior Deck Detail

2x6 STUDS @ 24" O/C
(UPPER LEVEL)

CONT. 1 1/4" TIMBER
STRAND RIM BOARD
OR APPROVED EQUAL

WOOD I JOISTS-
SEE PLAN

2x6 DECKING
(TRTD.)

DOUBLE 2x6 TOP PL (TYP)

DECK HT. PER
ARCH. REQMTS.

2x6 STUDS @ 16" O/C
(LOWER LEVEL)

DBL 2x10 (S.T.)
BLOCKS CONNECTED TO
STUDS EA. END

2x10s (S.T.)
AT 16" O/C

GALV. JOIST
HANGER (TYP)

CONT. DBL 2x10 (S.T.)
CONNECTED WITH 5/8",
GALV. THRU-BOLTS AT 16"

TYP. I–JOIST & EXTERIOR DECK DETAIL

NOT TO SCALE   (DETAIL T6–IJ5)

(By permission from The McGraw-Hill Co.,
Structural Details Manual, David R. Williams.)

FIGURE 17.15 I-joist connected to exterior wall with deck.

MEDIUM DENSITY FIBERBOARD (MDF)

This product rates a separate section inasmuch as it is widely used in the manufacture of furniture, cabinetry, door parts, millwork, and laminate flooring.

MDF is a composite material consisting typically of cellulosic fibers combined with a synthetic resin or other suitable bonding agents and joined together with heat under pressure. The resulting panels are flat, smooth, dense, and free from knots, grain, and other natural defects present in lumber. As such, MDF is the perfect material for any number of products:

- Kitchen cabinets: MDF wrapped with an overlay; solid color or wood grain or bonded to a printed design finds application in shelving, doors, and front and side panels.

- Paneling: frequently used as a core material for wood veneers; often printed

- Doors, jambs, and millwork: MDF is the perfect material for these applications since it is warp-resistant and has a smooth surface and has some insulating qualities; it can be cut into a wide variety of shapes and sizes, making it a popular medium for molding and millwork manufacturers

MDF does require different care and treatment from other products. It expands to a greater degree than does wood and therefore requires more attention to proper climatizing. When a decorative face or veneer is applied to one side, MDF's ability to absorb moisture makes it susceptible to warpage, and a backer sheet is mandatory.

When used as a substrate for plastic laminates, MDF or particleboard panels are subject to warpage if not stored in a properly climate-controlled area. When an unbalanced MDF laminate panel is produced with no backing sheet, warpage can also occur. When these unbalances continue, stresses within the panel become excessive, and cracks can appear in the laminate face. Unbalance can occur for any of the following reasons:

- When wood veneer is applied and not sealed
- Lack of conditioning of the substrate and the veneer face
- Product design problems
- The environment in which the panel and face are to be installed.
- Unusually moist or dry conditions in either storage or laminating areas

fastfacts

When installing MDF underlay, leave a ⅜-inch gap (9.5mm) at all walls. Arrange panels so that all four corners do not meet at one point. Butt all panel edges and ends to a light contact.

Wood veneers are applied to MDF and particleboard to create attractive paneling; vinyl films applied to a base MDF/particleboard substrate to create a wood veneer finish, at first glance, can fool even the most experienced carpenter.

One of the more common uses of MDF/particleboard is as an underlay for carpet or resilient flooring. When this material is used for that purpose and is to be nailed or stapled to the subfloor, the sublfloor should be at least ¹⁹⁄₃₂ inch (15.08mm) thick with a minimum of ³²⁄₁₆ panel span rating. When particleboard is to be glue-nailed, subfloors should be at least ¹⁵⁄₃₂-inch (11.9mm) plywood or ⁷⁄₁₆-inch-thick (11.11 mm) OSB with a minimum of 24/16 panel span rating.

18

GYPSUM DRYWALL, METAL FRAMING, AND SOUND AND MOISTURE CONTROL

S teel or wood stud framing, faced with gypsum drywall panels (regular, fire-rated, or vinyl-covered), have dominated the interior partition industry, replacing the costly and highly skilled method of constructing plaster walls These drywall partitions are lightweight, sturdy, cost-effective, and quick to complete.

Although wood stud framing appears to dominate the single-family- home industry, these builders are taking a hard look at metal stud framing—the mainstay in commercial and institutional construction. One of the problems with wood framing is shrinkage in the members, causing twists, warps, or bows. Even when straight wood studs are placed in a wall, prolonged exposure to damp and moist conditions may cause them to twist, warp, or bow. Replacing or correcting these members is essential if a true drywall surface is to be achieved. USG's (formerly known as United States Gypsum) framing manual (Figure 18.1) describes and illustrates several methods of correcting framing problems.

The advantages of steel framing over wood framing are many:

- Steel studs are straight and true, unlike the non-uniform characteristics of wood.

- Steel studs do not warp, split, twist, or shrink.

- Steel studs are impervious to termite infestation.

The moisture content of wood framing should be allowed to adjust as closely as possible to the level it will reach in service before gypsum drywall or plaster base application begins. After the building is enclosed, delay board application as long as possible (consistent with schedule requirements) to allow this moisture content adjustment to take place.

Framing should be designed to accommodate shrinkage in wide dimensional lumber such as is used for floor joists or headers. Gypsum wallboard and veneer plaster surfaces can buckle or crack if firmly anchored across the flat grain of these wide wood members as shrinkage occurs. With high uninterrupted walls, such as are a part of cathedral ceiling designs or in two-story stairwells, regular or modified balloon framing can minimize the problem.

Framing Corrections If joists are out of alignment, 2″ x 6″ leveling plates attached perpendicular to and across top of ceiling joists may be used. Toe-nailing into joists pulls framing into true horizontal alignment and ensures a smooth, level ceiling surface. Bowed or warped studs in non-load bearing partitions may be straightened by sawing the hollow sides at the middle of the bow and driving a wedge into the saw kerf until the stud is in line. Reinforcement of the stud is accomplished by securely nailing 1″ x 4″ wood strips or "scabs" on each side of the cut.

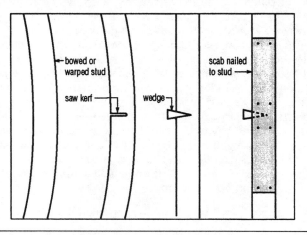

FIGURE 18.1 Correcting wood framing members.

- Steel studs can meet more stringent wind-loading requirements than wood.
- Steel truss members can span longer unsupported distances than wood, thereby opening up large areas in the home.
- Steel is a recyclable material; today's metal studs contain, on average, about 25% recycled material.
- There is less waste when using steel studs if sufficient quantities are ordered; they can be shipped pre-cut to the exact size required.

The main disadvantage of steel stud framing is the lack of familiarity with its installation. Metal cutters and snips replace power saws; powder-activated fasteners and screws replace nails. Engineering and design costs for metal stud framing are slightly higher than for wood framing. Carpenters trained and experienced in wood stud framing are often put off by metal stud framing, which presents a somewhat shaky partition until drywall is screwed on, turning the wall into a rigid diaphragm.

Carpenters also need to rethink the way in which they install blocking for kitchen cabinets, bath vanities, toilet accessories, or future heavy wall-mounted items. You can use 6-inch (15.24cm) to 10-inch-wide (25.4cm) 20-gauge galvanized metal strips spanning three or four metal studs instead of 2 x 4 (5.08cm x 10.16cm) or 2 x 6 (5.08cm x 25cm) wood blocking in wood stud framing.

HYBRID FRAMING

Just like testing the water by dipping your toe in a swimming pool, some builders are using hybrid framing—a combination of wood and steel stud members. Steel can be used for non-load-bearing members, while load bearing-walls are wood framed, or vice-versa.

fastfacts

You don't see nail pops when you frame with metal studs; how much it is worth to eliminate nail pops?

Steel trusses, supported by wood framing, can used for floor framing. As builders become more familiar with steel framing, they can gradually move toward an all steel-framed house.

METAL FRAMING

Metal stud framing consists of a limited number of different products:

- Steel studs: 1⅝ inch (4.127cm), 2½ inch (6.25cm), 3½ inch (8.75mm) .4inch (10.16cm) 6 inch (15cm), 8 inch (20 cm)

- Steel furring channels: also referred to as "high-hats" because their configuration looks like a man's hat; these channels are used primarily when drywall is applied on CMU walls.

- Steel joists: sometimes called super C joists, they function in the same way as 2" x 8" (5.08cm x 20cm), 2" x 10" (5.08cm x 25cm), or 2" x 12" (5.08cm x 30.48cm) wood floor and roof joists function.

Both load-bearing and non-load-bearing studs are available in a variety of metal gauges depending upon their function. Non-load-bearing partitions are generally 25-gauge studs, and load-bearing partitions are 20-gauge. However, various factors come into play during the selection of the proper stud: height of the wall framing, wind load, and anticipated dead or live loads.

Deflection

Many of the tables for metal stud selection contain a formula known as L over 360, expressed as L/360. An established engineering rule is that a framing member should not deflect (bend) more than 1/360 of the length of its span when the span is expressed in inches (will need to convert to metric, if applicable). For example, a 30-foot (99 meters) long-beam (360 inches) should not deflect more than 1 inch (2.54cm.) Other "L over" tables include L/240 and L/120, depending upon the engineering design.

Studs are available in heavier gauges—18, 16, 14—and are referred to as "structural studs." These framing members require engineering design input to determine spacing, height, dead and live loadings, and wind loadings. Structural studs are assembled by welding, unlike the screw fastening of lesser gauge (25, 22, 20 gauge) members.

GYPSUM WALLBOARD OR DRYWALL

Gypsum is a white-colored mineral, chemically known as calcium sulfate. After this mineral is mined, it is crushed, dried (it contains about 20% water), and ground to a powder-like consistency. After the water has been removed, this material becomes plaster of Paris. The wallboard manufacturers add various other ingredients to the plaster of Paris and sandwich it between two sheets of specially manufactured paper; it then emerges as gypsum wallboard.

There are various grades of drywall for both interior and exterior use:

- General-purpose gypsum drywall: thicknesses ranging from ¼" (6.35mm), ⅜" (9.5mm), ½" (1.25cm) to ⅝" (15.87mm) and a special board called Shaftwall™ that is 1-inch thick (2.54cm)
- Fire-code or fire-rated drywall with non-combustible core and paper facings: offers greater protection against fire; some manufacturers add glass fibers for added strength; Type C fire-code gyp board is superior to the standard "X" type board.
- Water- and moisture-resistant drywall: frequently referred to as "green board"—the color of the paper sandwich; there is also a fire-rated waterproof/moisture-resistant version.
- Gypsum board: used as a base for veneer plaster, referred to as "blue board" because of its bluish-gray face paper
- Sag-resistant drywall: especially made to hold up under high humidity conditions and wet- application texturing
- Soffit board: a weather-resistant board when installed in sheltered exterior surfaces such as car ports and open porches
- Abuse-resistant drywall: a fiberglass and heavy facing paper-reinforced product designed to resist or lessen damage due to impact, indentation, and abrasion of the surface, available in regular and fire-code gypsum core
- Sheathing: the outer layer on an exterior wall that will be covered by brick veneer, vinyl siding, or some other decorative wall finish; some exterior gyp board sheathings have a wax emulsion coating; this premium exterior sheathing product is easily recognized by its yellow color and is fiberglass-reinforced for superior durability
- Foil-backed: creates an interior wall surface and a vapor retarder in one product; for use on exterior wood or steel stud wall construction, furred masonry walls, it helps to prevent moisture from entering exterior wall and ceiling spaces

- Moisture- and mold-resistant gypsum panels: a relatively new product developed as a response to the growing concern about mold producing organisms

Gypsum drywall panels are available in standard 4-foot-wide (13 meter) sheets; 5-feet-wide (16/5 meter) panels are offered by some manufacturers. Panel lengths can be 8' (2.4 meters), 9' (2.7 meters), 10' (3.3 meters), 12' (3.6 meters), 14' (4.2 meters), and 16' (4.8 meters) long.

Drywall Finishing Procedures

As anyone who has ever attempted to tape drywall joints knows, the finishing of a drywall partition or ceiling largely depends upon the skill and taping techniques of the applicator. Depending upon the material or finish to be applied to the wall or ceiling, there are five levels of "finish" that can be selected:

1. Level 0: This level requires no taping or finishing and is often applied to temporary construction.

2. Level 1: All joints and interior angles are to have tape set in joint compound The surface shall be free of excess joint compound and tool marks; ridges are acceptable. All fastener heads are to be covered with taping compound. Frequently used in partition work above the ceiling to seal off areas and on walls that will be covered by paneling. Used in plenums above the ceiling or in concealed spaces Some degree of fire rating and sound attenuation can be achieved with this level.

3. Level 2: All joints, angles, fastener heads, and drywall accessories are taped with two different coats of taping compound applied over corner beads, expansion joints, etc. All surfaces are to be smooth and free of tool marks and ridges. Heavy textured finishes, paneling, and heavy vinyl wall coverings require this level of finish. Also used on garage walls, warehouse walls, and other areas where appearance is not the primary consideration.

4. Level 3: In addition to the level included in Level 2, an additional coat of joint compound, making a total of three separate coats, is needed to comply with this level of finishing. Surfaces must be free of all tool marks and sanded or troweled smooth to receive paint or wall coverings. Fastener heads and all accessories must receive three separate coats of joint compound.

This level is specified when flat paints, light textures, or wall coverings are to be installed. It is not recommended for gloss or even semi-gloss enamel paint finishes, if a truly seamless look is required

5. Level 4: In addition to the three coats required in Level 3, a thin skim coat of joint compound (or other material made expressly for this application) is applied over the entire surface. Fastener heads and accessories must be covered with three separate coats of joint compound. This level of finishing will approximate the look of plaster and is desirable when semi-gloss or high-gloss wall finishes are to be applied and where surfaces will be illuminated by wall washing-type light fixtures.

Note: Some manufacturers identify levels of finish differently, beginning with Level 1 and ending with Level 5.

Fire-Rated Partitions

Building codes specify fire ratings for drywall partitions and floor assemblies ranging from 1 hour to 4 hours (the time the partition or floor assembly must withstand the effects of a fire) depending upon their location and function. These ratings are for the purpose of delaying the spread of fire from one floor to another or from one room or building to another.

Sound-Rated Partitions

What is sound? You know it when you hear it, but what is it really? Sound is a vibration that occurs at various frequencies in an elastic medium—air, water, or gas. Sound pressure levels are represented in *decibels*—a ratio of intensity of sound to the threshold of hearing, designated by the symbol dB.

Decibel levels of common noises (approximate)

- Rustling of leaves   10 dB
- Inside bedroom, quiet conversation   30dB
- Private office conversation   40 dB
- General office conversation   50 dB
- Face-to-face conversation   60 dB
- Bathroom noise   70 dB

- Inside a speeding car   80 dB
- Stereo   90 dB
- Noisy party or symphony orchestra   120 dB
- Jet aircraft   140 db

STC Ratings

Sound-transmission-coefficient ratings are another means of defining noise levels. STC ratings apply only to those sounds that have the same frequency spectrum or sound profile as those produced by the human voice. An easy way to remember this is to think of STC as *Speech Transmission Class*. Various partitions are built to STC standards, which means that they will provide some protection against noise transmission. The higher the STC rating, the more effective the noise barrier.

Some common STC ratings:

- STC 25: normal speech that can be heard and clearly understood through a barrier
- STC 30: Loud speech that can be clearly overheard but not clearly understood
- STC 35: Loud speech that can be heard but is difficult to understand
- STC 42: Loud speech that can be heard but only faintly
- STC 45: Normal speech that cannot be heard
- STC 60: Loud speech that cannot be heard or other loud sounds that can barely be heard

Sound travels through the air—conversation, music, street noises—and through structures, footsteps on a hard surface or vibration from machines or equipment.

There are several ways to combat the transmission of sound:

- Mass: thicker wall and floor assemblies, thicker pad and carpet, lead lined sheet rock
- Decoupling: isolating the vibration-creating machines or appliances from the structure to which they are attached or set on
- Absorption: using materials that absorb sound such as fiberglass batts or blankets, sound-attenuation blankets, sound-absorption board

- Sealants: acoustical sealants or caulking materials to close off openings in walls, floors, or ceilings where electrical, plumbing, or ductwork penetrations exist

Dos and Don'ts for Sound Control

- Seal the bottom of the drywall by placing a heavy bead of acoustical caulking adjacent to the runner prior to installing the gypsum board.
- Offset electrical outlet penetrations between rooms—don't install back to back.
- Seal around plumbing pipe penetrations through the floor and wall.
- Apply acoustical sealant around al wall and floor duct penetrations.
- When suspended acoustical ceilings are installed, drape a fiberglass blanket over the drywall partition that abuts the laid-in ceiling to prevent sound from traveling from room to room.

Figure 18.2 illustrates some of the ways in which sound reduction can be built into a drywall partition.

Plastic waste lines, PVC and CPVC, although much less expensive to install, create more noise as liquids flow through them; cast-iron pipe, more expensive to install, is much quieter. When plastic waste lines are installed, wrapping them in a fiberglass batt or blanket will deaden the noise considerably.

Mildew and Mold

The newspapers in early 2000 were sprinkled with articles about people initiating lawsuits when they discovered mold in their homes. There were articles about multi-million-dollar awards against builders. This type of litigation became the latest class-action drama featuring lawyers, hygienists, and toxicologists in starring roles.

When a building is under construction, there are many ways in which moisture can enter. If concrete slabs have not been allowed to cure thoroughly, they will remain repositories for moisture long after the building is enclosed. Wall and roof openings may have temporary protection while awaiting skylights and windows, and the polyethylene coverings over these openings may become damaged, allowing rain to enter the structure. Roof installations with

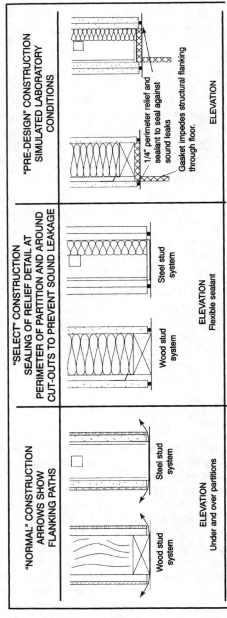

Sound Isolation Construction

"NORMAL" CONSTRUCTION
ARROWS SHOW FLANKING PATHS

Wood stud system

Steel stud system

ELEVATION
Under and over partitions

"SELECT" CONSTRUCTION
SEALING OF RELIEF DETAIL AT
PERIMETER OF PARTITION AND AROUND
CUT-OUTS TO PREVENT SOUND LEAKAGE

Wood stud system

Steel stud system

ELEVATION
Flexible sealant

"PRE-DESIGN" CONSTRUCTION
SIMULATED LABORATORY
CONDITIONS

1/4" perimeter relief and sealant to seal against sound leaks

Gasket impedes structural flanking through floor.

ELEVATION

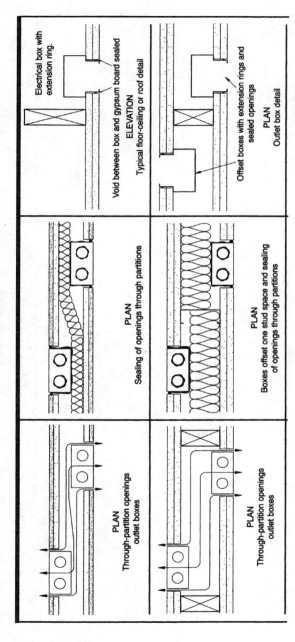

FIGURE 18.2 Sound reduction techniques in drywall partitions. *(continued on next page)*

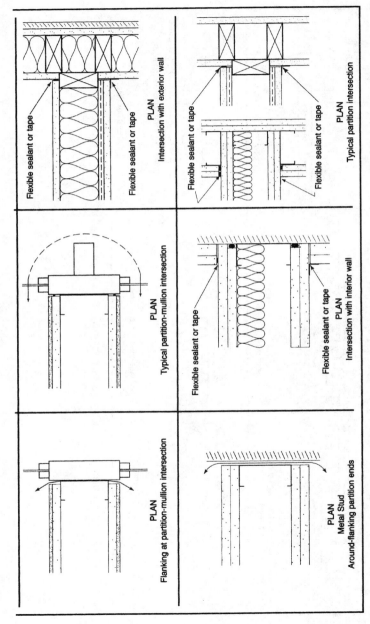

FIGURE 18.2 Sound reduction techniques in drywall partitions. *(continued)*

missing or yet-to-be-installed flashings are another source of water penetration, and it is not too uncommon to find drywall being installed in buildings and homes with little or no weather protection.

Trapped moisture in the home or building is one of the ingredients in the formation of mildew and mold. The other ingredient is one that provides food for the organisms to grow and flourish, and the paper face on sheets of drywall is an "all you can eat" incentive for mold to take hold and grow.

The following four conditions are necessary for mold formation and growth:

- Temperatures between 40 degrees F (5 degrees C) and 120 degrees F (49 degrees C).

- Spores: mold spores must be present to begin growth (these spores exist almost everywhere in the environment).

- Moisture: mold cannot extract moisture from the air; it must obtain it from a substrate.

- Nutrients: mold requires nutrients, and paper, dust, glue (a favorite food), and wood products provide them.

Mold grows and thrives best in a warm environment, between 77 degrees F to 86 degrees F (26 degrees C to 30 degrees C). The moisture it requires can come from leaks in the building or even from excessively humid air. Relative humidity above 70% is perfect for the formation of mold. Molds require oxygen but not light to survive, and if left unchecked, they will eventually eat away the organic material to which they have become attached.

When a house is under construction, even if the roof is tight, moisture in the form of condensation can still penetrate the walls. For example, rain falls on the surface of the exterior wall, the rain stops, the sun comes out, and the vapor pressure and dew point rise, in some cases creating condensation behind the sheetrocked walls, especially when there is no vapor barrier installed and no air circulation in the building.

Even lawn irrigation systems striking the outside wall may form mold "colonies." These colonies, either inside or outside, begin to grow in 1 or 2 days when damp materials are present. When the relative humidity lowers and the temperature rises or falls significantly, these colonies can become dormant. But when humidity rises again, the molds regenerate.

Health Hazards

The spores created by mold and mildew become airborne quite easily. If a furnace fan or air-conditioning fan is turned on when the house is under construction and mold and mildew are present, this untempered air will distribute spores throughout the home. If carpet has been installed and air is circulated under the same conditions, some spores will settle onto the carpet. Persons with allergies will experience cold-like symptoms when exposed to these airborne spores; some spores entering the lungs may cause serious respiratory illness. It is not healthy to live in a structure with high levels of mold and mildew spores.

How to Protect Your Home

- Fix leaks immediately; mold can develop within 24 hours after a water spill.
- Keep bathrooms, kitchens, laundry rooms, and basements well ventilated; open windows and turn on fans after bathing, cooking, or doing laundry.
- Inspect closets, bathrooms, basements, and other poorly ventilated areas for signs of mildew and mold; check corners, backs of closets, and out-of-the-way spaces carefully.
- Store textiles dry and clean.
- Prevent condensation problems by installing adequate insulation in walls; storm or insulated glass windows raise the temperature of the glass during winter months, resulting in less condensation on windows.
- Reduce sources of water coming into the house; check caulk joints annually and repair or replace as required; seal cracks in foundation walls; slope earth away from the foundation walls to obtain positive drainage of rain water.

fastfacts

Long-term storage of clothes or fabrics in plastic bags is not good: condensation can form in the bag. Desiccants such a silica gel can reduce moisture buildup.

fastfacts

Under the right conditions, mildew may have formed within the wall cavity of the sheetrock partition. Depending upon the extent of growth, the services of a industrial hygienist may be required to investigate and propose remedial action.

- Clean up small patches of mold by using a solution of 5 parts water, 1 part sodium hypochlorite (household bleach); another effective solution can be made from: ⅓ cup household detergent (powder or liquid), ⅔ cup trisodium phosphate (TSP, readily available in hardware and paint stores),1 quart 5% sodium hypochlorite solution (standard household bleach), 3 quarts of water; mix all ingredients thoroughly, then apply to the area where mildew or mold is present; wear protective clothing, rubber gloves, and eye protection; after 10 to 15 minutes rinse, allow to dry, and apply coat of stain kill prior to painting area.

Condensation

Undue amounts of condensation forming and remaining on the interior of a building can be a source of mold and mildew growth. A better understanding of the causes and cures for excessive condensation will help in avoiding another source of moisture that can contribute to the formation of mold and mildew.

The fogging up of windows in the winter indicates condensation excessive humidity inside the home. (When this phenomenon occurs outside the home, it is referred to as dew.) Excessive moisture in the home comes from cooking, running the shower, mopping the floor, and other everyday activities.

fastfacts

Did you know that daily living activities of a family of four can add as much as 18 gallons of water to the air in the house each week?

Condensation in a building may be the result of air infiltration. Small differences in pressure across the section of an exterior wall system can result in a large volume of moisture-laden air leaking into or out of a building, thereby increasing the potential for condensation buildup.

House or Building Wrap

The old method of stapling 15-pound felt to exterior walls has been largely replaced by materials referred to as ai- infiltration barriers. These materials are available in widths up to 9 feet (29.7 meters) allowing a builder to actually "wrap" a house in much the same manner as a Christmas gift, hence the name "house wrap."

The installation of the wrap, especially when large-width rolls are used, is very economical, but certain steps must be taken in order for it to operate at its peak efficiency. All splits, seams, and wall penetrations must be sealed with special adhesive-backed tape. Because air infiltration-barrier material retards the flow of moisture-laden air in and out of the building, it can be installed on either the inside or outside surface of the wall.

Figure 18.3 illustrates the proper installation of house wrap, using a header wrap and seam tape. Figure 18.4 illustrates the methods by which flashed window openings can be wrapped effectively with house wrap or building paper.

Relative Humidity

Air can hold only so much moisture, and temperature is a factor in how much it can hold. When air at a certain temperature contains all the vapor is can hold, it is said to have a relative humidity of 100%. When the air holds only 50% of the water vapor it can hold, it is said to have 50% relative humidity. Cooler air is capable of holding less humidity than warmer air so that air at 40 degrees F (5 degrees C) and 100% relative humidity will actually contain less moisture than air at 75 degrees F (24 degrees C) and 100% relative humidity.

New homes are much "tighter" than older homes, and when windows and doors are weather-stripped more efficiently and insulation with vapor barriers is installed, moisture tends to be locked up in the home.

Because the major model building codes allow air infiltration barriers to be used in lieu of building paper for most applications, the use of these products is on the rise. To get the full benefits of an air barrier, it must be sealed as described below to ensure that it is airtight.

Air infiltration barriers are available in rolls up to 9 ft wide, allowing the builder to wrap the barrier all the way around the house during construction. This is the origin of the term "house wrap." The large size speeds up installation and minimizes the number of seam seals. When the wrap is used as an air barrier, all of the splits, seams, penetrations and damaged areas must be repaired using a special adhesive-backed seam tape. In this respect, an air barrier differs from a vapor retarder.

Figures 16 and 17 show general installation techniques for proper application of an air barrier.

Vapor Transmission in Wood Wall Construction

Vapor transmission is the molecular passage of water through the components of a building. A differential water vapor pressure across the wall causes this movement. In cold weather, vapor from the interior of the structure can permeate through the interior wall finish and condense on cooler framing and sheathing surfaces in the wall cavity if there are surfaces colder than the dew point temperture. To prevent this, an effective interior ('warm side') vapor retarder, installed beneath the interior wall finish, is recommended by codes for most cold climate regions.

The "warm side" vapor retarder in exterior walls may be omitted in regions with moderate temperatures, such as the southern and southeastern United States. In warm, humid regions close to the Gulf of Mexico, and in Hawaii and the Caribbean regions, where air conditioning is prevalent, the vapor retarder should be installed on the exterior side of the wall, behind the sheathing. This will prevent humid air from penetrating into the wall cavity and causing increased condensation on the cooler interior wall surface.

FIGURE 18.3 Using housewrap with seam tape. *(Courtesy of LaFarge North America, continued on next three pages)*

Selection and Installation of Vapor Retarder

When the warm side is determined to be the inside wall, the vapor retarder can be a kraft paper or foil/kraft paper facing on the wall insulation. The effectiveness of this vapor retarder depends on how carefully the insulation is installed. The most effective installation technique is to cut the insulation batt length slightly oversize so it can be friction-fit to avoid gaps at the top and bottom wall plates. Also, the installation tabs of the insulation facing should be lapped and stapled onto the nailing surface of the studs, instead of the sides of the studs, to "seal" the insulation facing against air and moisture leakage, and to minimize gaps between the insulation and studs.

Alternatively, an effective continuous vapor retarder can be installed by using a separate layer of 4-mil polyethylene sheeting, stapled over the interior side of the wall framing. In this case, unfaced insulation without an integral vapor retarder facing may be used, and friction fit to fill the stud cavities without gaps.

While polyethylene sheeting makes a very good vapor retarder, it is relatively difficult to install. In most cases, the use of polyethylene is not necessary, even in very cold regions. Ordinary interior latex paint applied over drywall can provide sufficient vapor retardant properties.

FIGURE 17

AIR INSTALLATION DETAILS USING HEADER WRAP

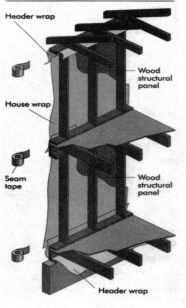

FIGURE 18.3 Using housewrap with seam tape. *(continued)*

MOISTURE FROM CONDENSATION

Condensation of vapor is a source of moisture intrusion. Condensation occurs if there is a significant drop in the air's temperature as it passes through an insulated wall such that the air temperature falls below the *dew point*. The dew point is the temperature at which moisture vapor in the air condenses. If it happens to be within the wall cavity, the building materials absorb this moisture, and thus the moisture content of the building materials increases. The moist air can enter from the inside or the outside, depending on the vapor pressure differential across the wall. In a hot, moist climate with air-conditioned buildings, there could be infiltration from the outside to the inside. In cold, dry climates, the inside air leaking out could cause the problem.

Air Infiltration in Wood Wall Construction

Condensation in wall systems may be caused by air infiltration. Even relatively small differential pressures across a given wall can cause a large volume of moisture-laden air to leak into or out of a structure, thereby increasing the risk of condensation within the wall.

Air Infiltration Barriers

Differential air pressures existing across the wall cause air infiltration. This differential air pressure can be caused by an unbalanced ventilation system, the stack effect caused by hot air rising within the structure, the use of unvented heating appliances, or wind. The actual differential pressure does not have to be very large to cause a significant amount of air leakage in one direction or another. If the moisture-laden airflow persists for a significant length of time, the moisture buildup can cause moisture damage to the structure and degrade the living conditions therein.

An air infiltration barrier such as house wrap retards the flow of moisture-laden air into the wall cavity. Because it does not matter where the airflow is stopped, the air barrier can be placed on the inside or outside surface of the wall. In a cold climate that requires a warm-side vapor retarder, the vapor retarder may act as the air barrier as well, if properly applied and sealed.

FIGURE 18.3 Using housewrap with seam tape. *(continued)*

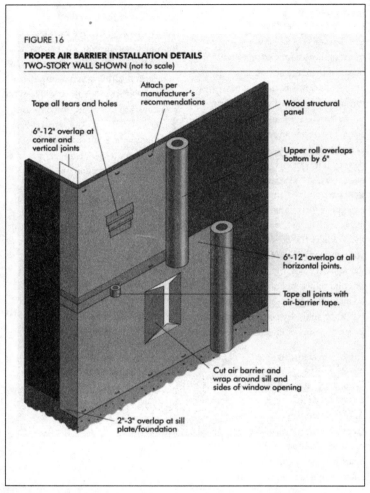

FIGURE 16

PROPER AIR BARRIER INSTALLATION DETAILS
TWO-STORY WALL SHOWN (not to scale)

Tape all tears and holes

Attach per manufacturer's recommendations

Wood structural panel

6"-12" overlap at corner and vertical joints

Upper roll overlaps bottom by 6"

6"-12" overlap at all horizontal joints.

Tape all joints with air-barrier tape.

Cut air barrier and wrap around sill and sides of window opening

2"-3" overlap at sill plate/foundation

FIGURE 18.3 Using housewrap with seam tape. *(continued)*

Water can accumulate in walls from two sources: water leaks, and vapor laden air that penetrates the wall to produce condensation. Water from leaks presents the greatest threat of water accumulation in walls. Since water can leak directly into the wall, it can quickly accumulate to levels that will degrade the wood components as well as other products in the wall. Moisture vapor from air penetration and vapor diffusion are important, but represent much smaller amounts of water accumulation.

HOW WATER LEAKS INTO WOOD WALL CONSTRUCTION

Water leaking through the envelope of a structure is the largest contributor to building damage. Leaks are caused by a number of factors, including:

- Improper or missing flashing
- Improper installation of weather-resistive barriers
- Poorly designed or executed wall intersections and penetrations

Wood structures have the ability to absorb, distribute and dissipate small amounts of water, especially from intermittent sources. The problems arise when there are design or construction errors that allow water into wall cavities at a rate that exceeds the structure's ability to absorb and eliminate the water. Wood construction will perform indefinitely but is subject to failure if exposed to prolonged wetting where the wood moisture content exceeds 19 percent.

FIGURE 18.4 Flashing in window openings. *(Courtesy of LaFrage North America, continued on next page)*

To reduce indoor humidity in the winter:

- Reduce the amount of moisture you put into the air; if there is a humidifier on your furnace, turn it down somewhat during cold weather.
- Vent all appliances outside: clothes driers, kitchen and bath exhaust fans
- Avoid storing wood inside the home

Higher humidity in the home in winter, particularly when heating is accomplished by a hot-air furnace, may actually provide more comfort for some people. So homeowners must weigh the advantages and disadvantages of high or low humidity levels.

To build a low-condensation home, follow these guidelines:

1. Choose wood or vinyl-sheathed wood windows rather than metal; this will reduce condensation on frames and sashes

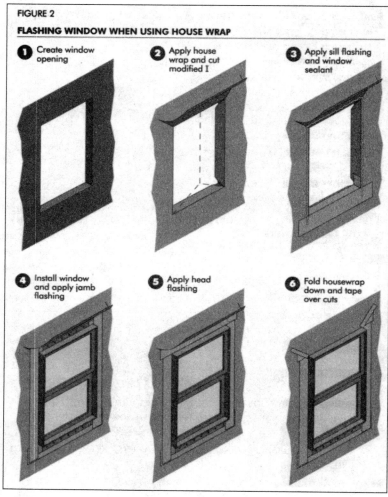

FIGURE 18.4 Flashing in window openings. *(continued)*

since wood is one of nature's best insulators. (You can purchase metal windows with a "thermal break" that actually separates the outer frame from the inner portion.)

2. Have your contractor use kiln-dried lumber and make sure he or she stores it covered and out of the elements.

3. When considering a crawl space or an attic space, make sure there is adequate cross-ventilation. Install a polyethylene vapor barrier over the crawl space.

4. If a forced-air furnace is the heating source, install an outside fresh-air intake on the system.

5. Provide outside vents for clothes dryers and all gas appliances, since water vapor is a by-product of gas combustion.

6. If you plan to have a basement, install a vapor barrier under the slab; apply a waterproofing mastic on portions of the foundation walls below grade; and allow for ventilation of the space.

7. Place heat vents below large glass areas, large window areas, and near patio doors

8. Install exhaust fans in kitchen, bathrooms, and laundry. Some kitchen hoods are known as "ductless," meaning that they merely recirculate air through a filter and don't exhaust it outside. They might be acceptable as a replacement in older homes without vents but should not be used in new construction.

9. Select windows and patio doors with double pane or insulated glass.

Rain

All the roofing, flashing, and air-infiltration work discussed to this point is based upon keeping water out of the home, easing the passage of vapor out of the house, and retarding the passage of moisture-laden air into the exterior wall cavity. But even the most dedicated efforts to reduce moisture infiltration often overlook the potential impact of wind-driven rain on the exterior wall of a structure. In some areas of the country high-wind-driven rain occurs more frequently than in other locations, and if your home is in one of these areas, there are other precautions that can be considered. Cavity wall construction is based upon the theory that water will penetrate the outer surface of an exterior wall and that the construction of the cavity must allow for this water to be collected and directed to the outside.

In areas where these wind-driven rains occur either frequently or infrequently, even with a well-constructed cavity wall system water can still find its way into the building envelope, particularly when it hits the wall at an angle. Gravity may need some help in dispelling water from a wood cavity wall, and this can be accomplished by utilizing what's called double wall construction (Figure 18.5). Pressure-treated furring strips, ¾ to 1 inch (1.90 cm to 2.54 cm) thick, are installed vertically as spacers, open at the bottom to allow any water intrusion to fall down and out; a bird or insect screen installed at the bottom opening will keep out unwelcome intruders.

Not only does this system allow for rapid dewatering, but the open cavity permits the equalization of pressure in the interior wall and the air space behind the cladding, acts as a barrier of sorts in reducing or eliminating the driving force of the rain from penetrating the skin of the structure.

INSTALLING RAIN-SCREEN WALLS

The entire exterior finish, weather-proofing, and flashing system in wood construction relies on gravity to keep bulk water out of the building envelope. Wind-driven rain can compromise these safeguards because the water is hitting the wall from a different angle. If wind-driven rain is an infrequent occurrence, the forgiving nature of wood construction can often account for the occasional influx of water into the building system. The moisture will be removed through capillary suction and the entire building frame will dry.

In areas where wind-driven rain is frequent, the amount of water driven into the wall system could be more damaging. In these cases, double-wall construction – also known as a rain-screen wall – is often used. Double-wall construction creates an air space between the exterior finish system and the weather-resistive system. This separation is made with the use of pressure-treated lumber spacers that are installed vertically and carefully detailed around openings and penetrations to allow drainage of any water that makes it through the exterior finish. This space – 3/4 to 1 inch – is open at the bottom to promote drainage and closed at the top to allow the air space to equalize with the exterior air pressure. The opening at the bottom has a pest screen. This system is often used with an interior air barrier to allow the air pressure in the interior of the wall to equalize with that in the air space behind the cladding. This will eliminate the driving force that causes water to leak into the wall.

FIGURE 18.5 Double wall construction when driving rain is a problem. *(Courtesy of LaFarge North America, continued on next two pages)*

FIGURE 18

RAIN-SCREEN WALL DETAILS

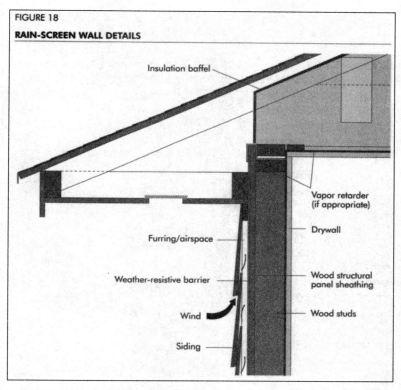

FIGURE 18.5 Double wall construction when driving rain is a problem. *(continued)*

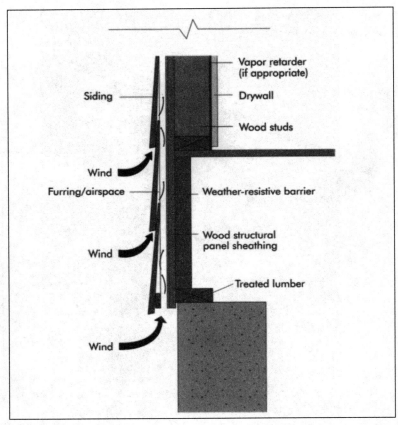

FIGURE 18.5 Double wall construction when driving rain is a problem. *(continued)*

chapter

19

FIREPLACES AND CHIMNEYS

ireplaces, wood stoves, flues, and chimneys can be part of any job. Some people want a wood stove installed to use as a backup heat source. Other people want them to use as an alternative heat source. Fireplaces are desirable in the eyes of many people. With either a fireplace or a wood stove, a chimney of some type is required. The chimney may be made of metal pipe or masonry materials. Fireplaces can be pre-fabricated metal units that don't require a lot of alteration in existing construction, or they can be masonry monsters that mandate a footing and substantial changes to existing construction conditions. Some contractors are not familiar with what is involved in the installation of a fireplace or wood stove. For this reason, we are going to go over the basics. Before we do, however, I want to stress that you should consult local authorities with regard to current local code and safety requirements before installing any type of fireplace or stove.

CHIMNEYS

Chimneys have been made with little more than sticks and mud. This type of construction is no longer acceptable. Today's code requirements for chimneys and flues are much more restrictive, and with good reason. Unlined brick chimneys often lead to unwanted chimney fires. Mud and sticks are not a prime choice in terms of chimney materials. But even after throwing out some of the more primitive methods for building a chimney, we are still left with options. Do you know what they are?

Masonry

Masonry chimneys are looked upon as being ideal. They are expensive, and they need to be cleaned periodically, but they are generally considered to be the best type of chimney available. I don't disagree with this view completely, but neither do I feel that a masonry chimney is always best.

When a masonry chimney is built, it must have a solid platform to rest on. This typically entails the use of a footing. Exterior masonry chimneys are usually covered in brick for appearance purposes, and this runs their cost up. If I were installing a wood stove, I probably would not opt for a masonry chimney. However, if a masonry fireplace is being built, it will be served by a masonry chimney. How involved is it for a general contractor to undertake the installation of a masonry chimney for a new home?

The first choice someone has to make is whether the chimney will be installed within the home or on an exterior wall. Interior chimneys can be enclosed with standard building materials, and this eliminates the need for a brick exterior, with the exception of where the chimney exits a roof. When brick is avoided, there is a substantial savings in cost.

Before a chimney can be installed inside a home, a proper space for it must be found. Local codes will set requirements for clear space around the chimney. When planning an interior chimney location, you must take the clearance requirements, the size of the flue and chimney, and the concealment framing all into account.

fastfacts

Costs for chimney construction can be kept down by having the chimney enclosed within the structure being served. A chimney that runs exposed on the exterior of a building requires more expense for the finished product.

TABLE 19.1 Modular clay flue lining sizes

Minimum net interior area (square inches)	Nominal dimensions (inches)	Outside dimensions (inches)
15	4 × 8	3.5 × 7.5
20	4 × 12	3.5 × 11.5
27	4 × 16	3.5 × 15.5
35	8 × 8	7.5 × 7.5
57	8 × 12	7.5 × 11.5
74	8 × 16	7.5 × 15.5
87	12 × 12	11.5 × 11.5

The amount of room needed for all this can be quite a bit. So much, in fact, that an interior chimney of this type may not be practical.

If a chimney is to be installed along an exterior wall outside a home, a footing and concrete pad will be needed for support. Existing siding on the home will have to be cut so that the chimney can be built along the sheathing of the house. Then flashing will have to be installed along the length of the chimney. All of this runs the expense of such a job up. Is your customer willing to pay thousands of dollars for a chimney? It could easily cost that much to build a masonry chimney. Is there an alternative? For wood stoves and prefabricated fireplaces there is.

Metal Chimneys

Metal chimneys are far from cheap, but they are much less expensive than their masonry cousins. Triple-wall metal chimneys can be installed with a minimum amount of clearance. This allows them to

TABLE 19.2 Recommended flue sizes for residential applications

1. Fireplaces: Flue should contain a minimum of 50 square inches and be no less than one-half the size of the fireplace opening.
2. Boilers and furnaces: Flue should contain a minimum of 70 square inches.
3. Room heaters and typical stoves: Flue should contain a minimum of 40 square inches.
4. Small stoves and heaters: Flue should contain a minimum of 28 square inches.

be installed in smaller spaces than a masonry unit could be. Almost anyone can install a metal chimney, so the high cost of a mason is not necessary. An average carpenter can easily install a metal chimney, from start to finish, in less than a day. This is much faster than the time it would take to have a masonry chimney built, and there are other advantages to metal chimneys.

Unless a customer wants a brick chimney for status or appearance, there is little reason to use a masonry chimney for anything other than a masonry fireplace. Metal chimneys can be installed inside or outside of homes. When installed outside, the chimney pipe can be enclosed with framing and siding. Inside, the pipe can be hidden with framing and drywall. Framing and siding can be installed around the chimney where it exits a roof.

The relative low cost of a metal chimney makes it not only a viable option but an affordable one. Since metal chimneys have a very smooth finish on the inside, they are not as prone to catch and hold creosote. This is not to say that they shouldn't be cleaned, but they may be less dangerous than a masonry chimney that has rough surfaces along its channel.

Few general contractors are accomplished masons. If you have a customer who wants a masonry chimney, you should have an experienced mason visit the job site with you before making any commitments. The mason will be able to point out alterations that will be needed on the existing construction. This helps to protect you, as well as making your estimate more accurate.

When a metal chimney is suitable, you may have your own crews take care of the installation. Suppliers of metal chimneys usually stock a variety of kits and accessories to make installations safer and easier. For example, you can get a through-the-wall kit or a through-

fastfacts

Many builders have their own people install metal chimneys. This is fine, but make sure that your people use the right materials, maintain the proper clearances, and install the material professionally. These types of chimneys are simple to install, but they do require a responsible approach to maintain a safe installation.

the ceiling kit. Each kit will contain special fittings that are designed to provide proper clearance and protection when a chimney pipe penetrates a wall or ceiling.

Let's say, for example, that you have a customer who wants a wood stove installed in a basement. The house has a basement and one level of living space. There is a standard gable attic over the living space. Your customer wants the chimney to run up through the house and has agreed to forfeit a small section of a room to house the chimney. What will you need to make this installation?

The first thing you will need are three collars that will be used where the chimney comes through the floor, ceiling, and roof. The collar will mount between joists, although you may have to cut out and head off enough space to accommodate it. The first collar will be installed in the basement. The exposed side of the collar will accept standard stove pipe. The upper side of the collar will accept the metal chimney pipe. This provides support for the chimney and complies with clearance requirements.

A section of chimney pipe is attached to the first collar, and subsequent sections are installed, one on top of the other, until the ceiling area is reached. At this point, another collar is needed to make the penetration through the ceiling and into the attic. Chimney pipe is installed on both sides of the collar and extended up through the attic. A hole is cut in the roof, and a third collar guides the chimney pipe through the roof. Once the pipe is out of the house, it is extended to meet local code requirements. A spark arrestor will normally be installed, as will a chimney cap. Other supports and accessories may be needed, depending upon individual conditions.

Once the chimney is installed and inspected, it can be concealed in a framed chase. Some clearance from combustible materials will be required by local regulations, but the distance will be minimal. This type of installation is fast, easy, safe, and affordable. It is hard to beat.

If the customer had wanted the chimney to extend up the exterior of the home, a through-the-wall kit could have been used. A special collar is used where the chimney penetrates the exterior wall, and a wall bracket provides support for the vertical chimney on the outside wall. The pipe is run up the side of the house, using special brackets to hold it in place. When the piping is complete, a wood frame can be built around the chimney, leaving required clearance, and covered with siding. Either of these types of installation is much simpler than what would be required for a masonry chimney.

FIREPLACES

There are three basic types of fireplaces that you may be asked to install. The most common, and most expensive, is a full masonry fireplace. A second type of fireplace is a pre-fabricated unit that is designed to be built into a wall. The third type of fireplace is a free-standing unit. There are pros and cons to each of these types of fireplaces, so let's discuss them.

Masonry Fireplaces

Masonry fireplaces are typically considered to be the most desirable. They can be made to fit almost any location nicely, are extremely durable, and their firebrick linings are safe. The biggest drawbacks to masonry fireplaces are their cost to install and the work involved with the installation.

The consideration many homeowners look at when assessing fireplaces is the effect they will have on the appraised value of their homes. From my discussions with real-estate appraisers, masonry fireplaces do very well when it comes time for an appraisal. While it is unlikely that excess equity will result from adding a masonry fireplace, it is likely that most of the cost will be returned in appraised value.

Built-In Fireplaces

Built-in fireplaces cost a fraction of what a masonry fireplace does. Prefab metal fireplaces can be installed in any room without exces-

TABLE 19.3   Firebrick sizes

Brick type	Dimensions (inches)
9" straight	9 × 4½ × 2½
9" small	9 × 3½ × 2½
Split brick	9 × 4½ × 1¼
2" brick	9 × 4½ × 2
Soap	9 × 2¼ × 2½
Checker	9 × 2¾ × 2¼

TABLE 19.4 Number of firebricks needed
per square foot of coverage

Brick type	Laid flat	Laid on edge
9" straight	6.5	3.5
9" small	6.5	4.5
Split brick	13	3.5
2" brick	8	3.5
Soap	6.5	7
Checker	6	6

sive alteration. Installation is simple. The unit is set in place, metal
chimney pipe is run, and a wall is framed up around the fireplace
and chimney. These working fireplaces can add a touch of romance
to a master bedroom or warmth to a family room.

What is the major drawback to a metal fireplace? Well, it isn't
cost, because they are cheap. I've had both masonry and metal fireplaces in homes where I have lived. Without question, I have preferred masonry fireplaces. My personal experience as a user of a
metal fireplace is that the firebox is too small. The space is typically
short and narrow, which restricts log length and burning time
between trips to the woodpile. Other than this one complaint, I
don't know of any other serious drawback.

Freestanding Fireplaces

Freestanding fireplaces were very popular for awhile, but the infatu-
ation with them seems to have waned. These are units that are
intended to set out in the floor of a room. A chimney pipe extends
off the top of the unit and is usually left out in open view. These fireplaces are inexpensive, in relative terms, and they are easy to install.
Some are very attractive and quite functional. The freestanding
aspect makes these fireplaces more like a wood stove than a fireplace in terms of heating capabilities.

One potentially dangerous drawback to a freestanding unit is the
risk that someone will be burned. This is especially true when small
children are found in the vicinity. The relative low cost of freestanding and prefab fireplaces make up for the fact that neither of these
units do well in terms of appraised value.

TABLE 19.5 Clearances for wood stoves without heat shields

> - The back of a stove pipe should be at least 18 inches from any combustible material.
> - The back of a stove should be at least 36 inches from any combustible material.
> - The side of a stove should be at least 36 inches from any combustible material.
> - A distance of at least 18 inches should be maintained with some sort of fire-resistant base for the stove to sit on.

WOOD STOVES

Wood stoves don't require as much work to install as most fireplaces do. A chimney is built and the wood stove is set in place. A stovepipe connects the stove to the chimney collar, and the job is done. There are, of course, a lot of wood stoves to choose from on the market. Making a decision on which stove to buy may very well be the toughest part of the job. Fortunately, your customers won't expect you to hold their hand while they shop for a stove. As long as you are aware of local code requirements and the methods for installing a safe, acceptable chimney, you are in the clear on wood stoves.

If your customer is planning to have you install a stove, you might suggest a brick hearth and perhaps even a brick heat shield for the wall behind or on either side of the stove. When a suitable heat shield is used, a stove can be safely set closer to a wall. This conserves floor space for your customer. The heat shield may be made of metal, brick, tile, or some other approved material.

TABLE 19.6 Clearances for wood stoves with heat shields in rear

> - The back of a stove pipe should be at least 9 inches from any combustible material.
> - The back of a stove should be at least 18 inches from any combustible material.
> - The side of a stove should be at least 36 inches from any combustible material.
> - A distance of at least 18 inches should be maintained with some sort of fire-resistant base for the stove to sit on.

OTHER CONSIDERATIONS

Other considerations may come up when talking with customers about wood stoves and fireplaces. For example, your customer may want gas piping run for a gas log. In most jurisdictions, a licensed gas fitter will be required for such an installation. Many licensed plumbers are also licensed gas fitters.

You should strive to stay informed, but don't be afraid to ask a customer to allow you to research a subject. If you are asked a question that you are unable to answer with authority, don't bluff the homeowner. Request some time to look into the matter and then present your findings at a later date. It is better to say nothing than to say something that is not correct.

20

PAINTING, WALL COVERINGS, AND FLOORING

Today's painting materials have come a long way from oil-lead-based varieties. From water-based paints to specialized high-performance coatings, there is a paint for every need on the market.

The first water-based paints contained styrene or styrene butadiene and were known as *latex paints,* meant for interior use only. Over the years acrylic and acrylic ester resins were developed for both interior and exterior paint applications.

Water-based coatings have high permeability to water vapor, making them suitable for application on moist, porous surfaces such as wood, concrete, and masonry. The wide variety of coatings now available takes into account ease of application, long life, and environmental concerns. Let's take a look at some of the paints available today.

PAINT VARIETIES

Alkyd Paints

Alkyd coatings are produced by reacting a drying oil with an alcohol. As the solvent evaporates, the surface begins to dry and the resin base begins to cure. The more oil in the formulation, the longer the curing time takes. Alkyds are used in interior and exterior enamels and produce a durable finish. Alkyd coatings should not be applied directly on masonry or galvanized steel without a prior application of an alkali-resistant primer or sealer.

Latex Paints

Latex paints are formulated from synthetic acrylic or vinyl acrylic resins and pigments dispersed in water that contains a surfactant (a chemical that makes water "wetter"). Latex paints dry by evaporation; as water evaporates from the film, solvents allow the particles of resin to fuse together and form a continuous surface. Latex paint has excellent adhesion capabilities, a tough surface film, good color and gloss retention, and ease of application and clean-up. Most latex paints have to be protected from freezing and can't be applied in temperatures lower than 50 degrees F (10 degrees C)

Silicon Alkyds

These coatings begin life as alkyds but are modified to contain up to 30% silicone resins for greatly improved color and gloss retention. Silicon alkyds find application in coastal areas and in environments with intense sunlight.

Urethanes and Polyurethanes

Urethane and polyurethane coatings contain an isocyanate complex that forms a tough, hard, but flexible chemically-resistant surface. Urethanes are light-stable, gloss-retentive, and non-yellowing. These coatings find application on hardwood flooring and as protective coating on stained wood trim and cabinets.

Epoxies

Catalyzed epoxy paint is produced by combining epoxy resin with a curing agent. The mixture has a limited time of workability, referred to as "pot life," that can vary from minutes to hours. These coatings provide high resistance to abrasion when applied to wall surfaces and can be used on floors in high traffic areas. Most epoxies develop a non-progressive chalk face when used as an exterior finish but otherwise exhibit great durability. Water-based acrylic epoxies, while not approaching the durability and performance of their solvent-based cousins, are still substantial coatings and offer the advantage of low odor.

Epoxy ester coatings are made by reacting a drying oil with an epoxy resin; this material dries by solvent evaporation and cures by oxidation. Although not as hard nor as chemically resistant as catalyzed epoxies, these ester-based materials are still tough and are easier to apply. Unlike the catalyzed epoxies, these materials come in a single package and don't require a hardener in order to cure.

Zinc-Rich Primers

Zinc-rich primers contain at least 80% by volume of zinc particles. When applied to a steel substrate, zinc-rich primers will actually "corrode" to protect the steel, somewhat like the process of hot-dipped galvanizing. When applying zinc-rich primers to steel, the substrate must be thoroughly cleaned and present a slightly roughened surface.

Zinc-rich primers are available in two forms: organic, containing carbon, and inorganic, containing an inorganic zinc-silicate matrix.

Specialty Coatings

There are also many specialty coatings, such as intumescent (fire-retardant) paint, reflective coatings, and bitumastic-based materials for heavy-duty immersion use such as below-grade service.

No matter which paint or coating is selected, the surface to which it will be applied must be prepared properly.

SURFACE PREPARATION

Each surface (substrate) requires different preparation, but some procedures are common to all.

- Remove any surface mildew by washing with a solution of 1 part household bleach to 3 parts water; apply, scrub, rinse, and allow surface to dry.
- Wear protective glasses when preparing the surface, whether it be washing with a chemical, sanding, or wire-wheel abrading.
- Do not apply an exterior coating immediately after rain, during foggy weather, or when rain is predicted.

- Do not apply a coating when the temperature falls below 50 degrees F (10 degrees C).

Next we'll talk about specific surface preparation guidelines.

Aluminum

Remove oil, grease, dirt, or oxide (the rough surface caused by prolonged exposure to weather that appears on aluminum that has not been anodized) by cleaning with a solvent. Solvent cleaning will not remove mill scale or oxidation—you will need some abrasive action for that. Change the cleaning cloth regularly when wiping down with a solvent to avoid respreading oils and grease over previously wiped areas.

Asbestos Siding

Yes, some older homes still have transite (asbestos bearing) siding! Remove all dust, dirt, and grease. If the siding is weathered (which it probably is) and porous, it should be treated with a masonry conditioner before painting.

Concrete Masonry Units (CMUs)

Concrete or cinder-block walls should be inspected to remove and replace, if necessary, all loose mortar joints. The surface must be free of dust, dirt, and grease. The blocks themselves and the mortar should be allowed to cure at least 30 days at temperatures above 75 degrees F (24 degrees C) before applying a coating.

Brick

Preparation is similar to that for CMUs above, except that brick walls should be allowed to weather at least one year. The walls should then be wire-brushed to remove any efflorescence (that white stuff!) and coated with a masonry conditioner prior to application of a finish coat.

Concrete

Concrete should be cured properly and must be free of all surface contaminants—curing compounds, laitance (that thin cement-colored flaky topping sometimes found on portions of concrete slabs), form-release agents, loosely adhering particles, and dust. Older surfaces may require a light sandblast. These surfaces are best treated with a primer before final coat application.

Copper

Remove all oil, grease, dirt, and oxide by either hand- or power-tool cleaning.

Galvanized Metal

Allow galvanized metal to weather at least 12 months before applying a coating. Solvent-clean and then apply a primer specifically made for galvanized metal. If adequate weathering is not possible, apply a prime and finish coat to a test area and allow to dry for one week before checking for adhesion.

Drywall

All surfaces must be clean and dry, and nail or screw heads coated with taping compound. Spackled nail or screw heads and all taped joints should be sanded and free of dust prior to applying a prime coat. Exterior surfaces must be spackled with an exterior grade taping compound.

Hardboard and Composition Board

Some of these materials exhibit a waxy surface that must be removed with a solvent prior to coating. Exterior hardboard siding, whether primed or unprimed, must be thoroughly cleaned and primed with an alkyd primer.

Plaster

Newly plastered walls must be allowed to dry at least 30 days prior to painting. Rooms should be ventilated and heated in cold weather. Bare plaster must be hard before painting, and textured, soft, porous, or powdery plaster should be treated with a solution of 1 pint of vinegar to 1 gallon of water; this treatment should be repeated until the surface is hard; then rinse with clear water and allow it to dry.

CLEANING TECHNIQUES

The Society for Protective Coatings is a non-profit association devoted to users and providers of coating-related systems, primarily dealing with metal applications. It provides effective, commonsense standards and specifications for a variety of surface preparations, designated by SSPC numbers.

SSPC-SP1: Solvent Cleaning

Solvent cleaning will remove all visible oil, grease, soil, drawing and cutting compounds, and other soluble contaminants. Solvent cleaning will not remove rust or mill scale from metals. When utilizing SSPC-SP1 cleaning, rags or cleaning clothes should be changed frequently to avoid re-spreading dissolved contaminants over previously cleaned surfaces. Adequate ventilation must be provided when using this form of substrate cleaning.

SSPC-SP2: Hand Tool Cleaning

Hand tool cleaning removes loose mill scale, loose rust, and other foreign matter. It is not meant to remove hardened mill scale or rust or previously painted areas. Before hand tool cleaning, SSPC-SP1 cleaning needs to be employed.

SSPC-SP3: Power Tool Cleaning

This method of cleaning involves the use of power tools to remove loose mill scale, loose rust, and other foreign matter. It is not meant to remove inherent mill scale or rust or previously painted areas. Before power tool cleaning is employed, SSPC-SP1 cleaning needs to take place.

SSPC-SP4: White Metal Blast Cleaning

Prior to this type of cleaning, the surface, when viewed without magnification, must be free from all visible oil, grease, dirt, dust, mill scale, rust, paint, oxides, corrosion products, and other foreign matter. The SSPC-SP1 cleaning process must be done first.

The term "White Metal" refers to the appearance of steel after it has been sand blasted of its outer oxidized coating. The metal appears, if not white, then a silvery color.

Note: There is no SSPC-SP5 system.

SSPC-SP6: Commercial Blast Cleaning

This method incorporates the parameters of SSPC-SP4 but also leaves the surface clean of all matter except staining. Staining must be limited to no more than 33% of each square inch (6.452 square centimeters) of surface area and may consist of light shadows, slight streaks, or minor discolorations caused by stains of rust, mill scale, or previous applications of paint. Before blast cleaning, proceed with procedures outlined in SSPC-SP1(solvent cleaning).

SSPC-SP7: Brush-Off Blast Cleaning

A brush-off blast-cleaned surface, when viewed without magnification, should be free from all visible oil, grease, dirt, dust, loose mill scale, loose rust, and loose paint. Tightly adhered mill scale, rust, and paint may remain on the surface. Again, use SSPC-SP1 (solvent cleaning) prior to employing this cleaning method.

SSPC-SP8: Pickling

Complete removal of rust and mill scale by acid pickling, duplex pickling, or electrolytic pickling is used in this cleaning process. Note: There is no SSPC-SP 9 system.

SSPC-SP10: Near White Blast Cleaning

A surface cleaned by this method, when viewed without magnification, shall be free of all visible oil, grease, dirt, dust, mill scale, rust, paint oxide, and corrosion products. Staining should be limited to no more than 5% of each square inch (6.452 square centimeters) of surface area and may consist of light shadows, slight streaks, or minor discolorations caused by stains of rust, mill scale, or previously applied paint. Before blasting apply SSPC-SP1 cleaning.

SSPC-SP11: Power Tool Cleaning to Bare Metal

Metallic surfaces prepared to this standard, when viewed without magnification, shall be free of all visible oil, grease, dirt, dust, mill scale, rust paint, oxide, corrosion products, and other foreign matter. Slight residue of rust and paint may be left on the lower portions of pits if the original surface was pitted. Use SSPC-SP1 (solvent cleaning) cleaning first.

SURFACE CONDITIONS

Alligatoring and Wrinkling

This condition is caused by excessive build-up of layers of paint and seasonal temperature changes that cause the various layers to expand and contract, resulting in loss of adhesion. To correct this situation, power- or hand-sand or apply paint stripper to remove all layers of paint down to the substrate. Bare wood that has weathered should be sanded down to fresh wood. When bare wood is left exposed to the elements for more than one week, peeling of the new coat may occur, so apply a primer as soon as the wood surface has been properly prepared.

Caution: When power sanding wear eye protection; when applying a paint stripper be sure to do it with adequate ventilation.

Blistering and Peeling

Blistering and peeling can occur when moisture is trapped behind the painted surface or when the surface preparation was inadequate. Application of latex paint at temperatures below 50 degrees F (10 degrees C) will also result in peeling paint.

To correct this situation, remove loose paint with pressurized water and scrape any areas of loose paint that resist power washing. When moisture behind the paint is the main contributor to the peeling, caulk all cracks and holes in the substrate and install louvers or vents if there is humidity build-up in the room behind the painted surface. Scrape off paint from the problem area, sand to fresh wood, and feather the edge of any tightly adhered paint.

Note: If the substrate is hardboard or composition board, careful sanding is required to avoid damaging the surface.

Flaking

Flaking, or lifting of paint from its substrate in the form of flakes, generally occurs after the painted surface has cracked or checked, and this condition is often found on wood with flat, hard-grain patterns. Highly pigmented paints tend to flake when applied to these hard surfaces, usually because of poor penetration into the surface.

By lightly sanding to fresh wood, followed by a coat of quality undercoat material, this condition can be avoided. The wood must not be allowed to weather before repainting.

Burning

Stucco and masonry walls constructed with mortar joints contain hydrated lime, a material added to stucco and mortar mix to improve its workability. The alkalinity in mortar and stucco, generally in the range of pH 13 to 14 (neutral is pH 9), requires at least 30 days to drop to acceptable levels.

When these surfaces have "cured," they must be protected against any moisture intrusion, which will activate any remaining alkalinity in the substrate, causing efflorescence (that white fluffy stuff) to form on the surface of the masonry.

Both new and old masonry surfaces must be dry and clean of any surface contamination before painting. A 100% acrylic coating is the best paint to apply to these surfaces.

Peeling on Galvanized Metal

Paint has a difficult time adhering to smooth metal surfaces, and galvanized metal is also coated with a layer of oil to prevent white rust. Alkyd paints applied to galvanized metal will look fine for a while, but the zinc in the galvanizing process will eventually react with the alkyd binder and the paint will begin to peel.

In order to obtain a good paint job, the protective layer of oil must be removed from the galvanized metal surface. Solvent wiping with a water-soluble cleaning agent or naphtha will do, but cleaning with mineral spirits will leave a film residue that will lead to adhesion failure. Weathered galvanized metal develops a layer of "white rust" that must be removed, generally by wire brush, in order to obtain a good bond with paint.

INTERIOR AND EXTERIOR SURFACES

Your paint supplier will be able to provide you with specific paint recommendations depending upon your individual application. These general guidelines may also be of help (see Figure 20.1).

Exteriors

Wood

Paint: gloss finish
Alkyd primer—one coat
Alkyd gloss—two coats

Paint: semi-gloss finish
Alkyd primer—one coat
Acrylic latex house and trim—two coats

Paint: flat finish
Alkyd primer—one coat
Acrylic house paint—two coats

Wood: transparent finish
If new wood, allow to dry and then apply one coat of the Flood
Company's Seasonite™ or equal. After Seasonite™ has weathered
for about 9 to 12 months, clean with wood cleaner and apply
two coats of clear wood finish with ultraviolet resistant additive.

Wood: semi-transparent finish
Flat finish, exterior oil semi-transparent stain—two coats

Wood: solid stain
Oil finish, oil base solid hide stain—two coats
Latex finish, acrylic house paint primer—one coat, followed by
exterior acrylic solid hide latex paint—one coat

Concrete Block

Concrete block: flat finish
Block filler—one coat, acrylic flat house paint—two coats

Concrete block: semi-gloss finish
Block filler—one coat, acrylic semi-gloss house and trim paint—two
coats

Metals

Steel: unprimed
Rust-inhibitive metal primer or damp-proof red oxide metal
    primer—one coat. Alkyd gloss enamel, urethane modified—two
    coats

Steel: shop-primed
Alkyd metal primer or damp-proof red oxide metal primer—one
    coat
Alkyd gloss enamel—two coats

Steel: galvanized
Galvanized metal primer—one coat
Alkyd gloss enamel, urethane modified—two coats

Copper
Zinc chromate primer—one coat
Alkyd gloss enamel, urethane-modified—one coat

Interiors

Wood

Painted: flat finish
Alkyd primer-sealer—one coat
Latex wall paint, vinyl plastic flat—one coat

Painted: semi-gloss finish
Alkyd primer-sealer—one coat
Latex semi-gloss or vinyl acrylic semi-gloss enamel—two coats

Painted: gloss finish
Alkyd primer-sealer—one coat
Alkyd gloss enamel, urethane-modified—two coats

Wood: transparent finish
Oil-penetrating stain—one coat
Sanding wood sealer—one coat
Polyurethane varnish—two coats

Note: This system is not recommended for wood floors.

Wood: rough surface, natural flat finish
Stained: oil penetrating stain—one coat
Natural: clear wood preserver, penetrating oil wood stain
(untinted) —two coats

Modern technology has developed methods of treating certain species to extend their life when exposed to the elements. All lumber species used for exterior architectural woodwork, except species listed as "Resistant or very resistant" in the following tables (although it is desirable for those species) shall be treated with an industry tested and accepted formulation containing 3-iodo-2-propynyl butyl carbamate (IPBC) as its active ingredient according to manufacturer's directions.

Some domestic woods according to heartwood decay resistance:

Resistant or very resistant	Moderately resistant	Slightly or nonresistant
Cedars	Baldcypress (young growth) *	Ashes
Cherry, black	Douglas-fir	Basswood
Junipers	Pine, Eastern White *	Beech
White Oak	Pine, So. Longleaf *	Birches
Redwood, clear heart	Pine, Slash	Butternut
Walnut, black		Hemlocks
		Hickories
		Red Oak
		Pines (other than slash, longleaf, and E. white)
		Poplars
		Spruces
		True firs (western and eastern)

* - The southern and eastern pines and baldcypress are now largely second growth with a large proportion of sapwood. Substantial quantities of heartwood lumber of these species are not available.

Some imported woods according to heartwood decay resistance:

Resistant or very resistant	Moderately resistant	Slightly or nonresistant
Mahogany, American (Honduras)	Avodire	Obeche
Meranti **	European walnut	Mahogany, Philippine:
Teak	Mahogany, Philippine:	Mayapis
	Almon	White lauan
	Bagtikan	
	Red Lauan	
	Tangile	
	Sapele	

** - More than one species included, some of which may vary in resistance from that indicated.

DATA: U.S. Dept. of Agriculture, Forest Products Laboratory

FIGURE 20.1   Treatment for exterior woodwork.

fastfacts

Concerned that white paint may yellow as it ages? Stir in a drop of black paint.

Plaster and Gypsum Drywall

Latex flat finish, latex sealer-primer—one coat
Latex flat vinyl acrylic—two coats
Latex semi-gloss, latex primer-sealer—one coat
Latex semi-gloss, interior vinyl acrylic enamel—two coats
Latex gloss finish, acrylic latex undercoat—one coat
Alkyd gloss enamel, urethane-modified—two coats
Latex eggshell finish, latex undercoat—one coat
Acrylic eggshell finish enamel—two coats

EXTERIOR COATINGS

Many exterior metal products—doors, gutters, downspouts, and roofing—are painted with a high performance paint containing fluorocarbons. These products come with a 20-year guarantee, but they are not maintenance-free unless some procedures are followed to reduce the effect of airborne pollutants and atmospheric weathering.

These 20-year fluorocarbon paint coatings used primarily on exterior aluminum products are made with high-molecular-weight polymers that have been formulated into a dispersion coating for application at the factory. Polyvinylidene fluoride (PVF2) is the base ingredient in these coatings. Other high-performance coatings use siliconized acrylics, siliconized polyesters, and other synthetic polymers.

fastfacts

Painting your ceiling and want to keep paint from dripping down? Slide a paper plate up the handle.

Degradation of the paint surface is caused by the collection of airborne dirt and chemical pollutants in the presence of moisture, which increases the potential for erosion, corrosion, loss of surface gloss, staining, and discoloration.

Ultraviolet rays degrade the resin vehicle and color in the coating, resulting in a loss of gloss and the formation of powder on the surface. This powder is referred to as *chalking*, a change in both the appearance and color of the coating. Regular maintenance can prevent chalking.

Maintenance of products with this high-performance coating should begin as soon as possible after installation so as to remove dirt and/or pollutants.

The run-down from sealants can contribute to aluminum staining with these high-performance coatings. The oils and plasticizers in many caulking materials can bleed onto adjacent surfaces and cause stains and discoloration.

To learn more about maintaining high-performance coated aluminum products, contact the manufacturer or the American Architectural Manufacturers Association for cleaning and maintenance instructions.

PAINTING

How Much Paint Do I Need?

One way to determine how much paint is required is to convert percent solids by volume in a gallon of paint to area coverage when you know the wet mil thickness. These two elements, wet mil thickness and percent solids, are generally listed on each can of paint.

Some irregular surfaces may be difficult to estimate when determining how much paint to buy. Figure 20.2 contains formulas to convert square, round, spherical, cylindrical, and cone-shaped surfaces to square feet, and Figure 20.3 will allow you to determine square-foot areas for irregular surfaces such as corrugated metal siding or roofs and chain link fencing. And lastly, Figure 20.4 shows how to estimate the square footage per lineal foot of pipe and also includes some metric conversion factors when figuring areas.

Some Painting Tips

- Latex paint dries a couple of shades darker, and alkyd paints dry a half to one full shade darker than the color you first see when the paint is applied.

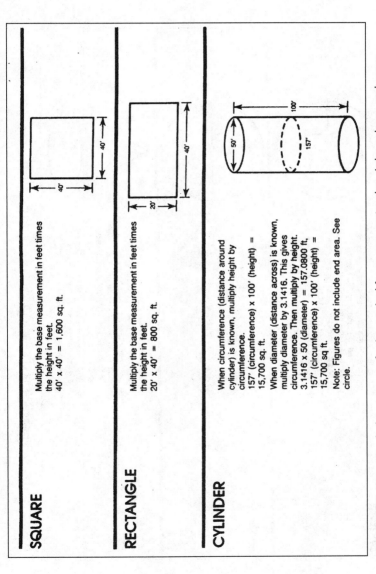

SQUARE

Multiply the base measurement in feet times the height in feet.
40' x 40' = 1,600 sq. ft.

RECTANGLE

Multiply the base measurement in feet times the height in feet.
20' x 40' = 800 sq. ft.

CYLINDER

When circumference (distance around cylinder) is known, multiply height by circumference.
157' (circumference) x 100' (height) = 15,700 sq. ft.

When diameter (distance across) is known, multiply diameter by 3.1416. This gives circumference. Then multiply by height.
3.1416 x 50 (diameter) = 157.0800 ft,
157' (circumference) x 100' (height) = 15,700 sq ft.

Note: Figures do not include end area. See circle.

FIGURE 20.2 How to calculate amount of paint required for squares, rectangles, circles, spheres, and cones. *(continued on next page)*

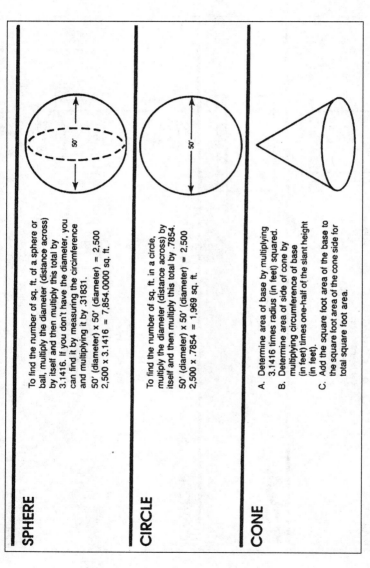

SPHERE

To find the number of sq. ft. of a sphere or ball, multiply the diameter (distance across) by itself and then multiply this total by 3.1416. If you don't have the diameter, you can find it by measuring the circumference and multiplying it by .31831.

50' (diameter) x 50' (diameter) = 2,500
2,500 x 3.1416 = 7,854.0000 sq. ft.

CIRCLE

To find the number of sq. ft. in a circle, multiply the diameter (distance across) by itself and then multiply this total by .7854.

50' (diameter) x 50' (diameter) = 2,500
2,500 x .7854 = 1,969 sq. ft.

CONE

A. Determine area of base by multiplying 3.1416 times radius (in feet) squared.
B. Determine area of side of cone by multiplying circumference of base (in feet) times one-half of the slant height (in feet).
C. Add the square foot area of the base to the square foot area of the cone side for total square foot area.

FIGURE 20.2   How to calculate amount of paint required for squares, rectangles, circles, spheres, and cones. *(continued)*

CORRUGATED SURFACES

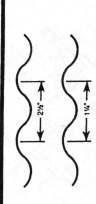

2½" Corrugated Sheet—To find width before corrugation multiply the width after corrugation by 1.08. Assume depth to be ⅝".

1¼" Corrugated Sheet—To find width before corrugation multiply the width after corrugation by 1.11. Assume depth to be ⅜".

ROOF DECK, METAL SHEETING

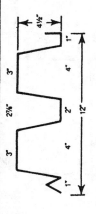

If the surface has a cross-section view similar to that shown, first figure the square foot area then multiply by 2.42 to obtain the actual surface area.

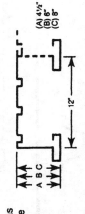

If the surface has a cross-section view similar to that shown, figure the top side as just the square foot area of surface. Figure the underside as follows:

A. For each square foot area multiply by 1.63 for actual surface area.
B. Multiply by 1.75.
C. Multiply by 1.92.

FIGURE 20.3 How to calculate amount of paint needed for corrugated surfaces and chain link fencing. *(continued on next page)*

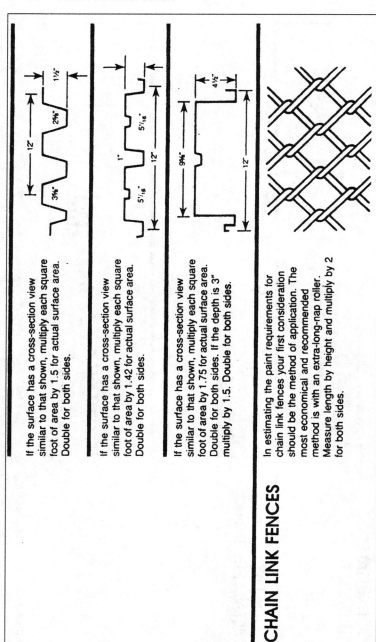

If the surface has a cross-section view similar to that shown, multiply each square foot of area by 1.5 for actual surface area. Double for both sides.

If the surface has a cross-section view similar to that shown, multiply each square foot of area by 1.42 for actual surface area. Double for both sides.

If the surface has a cross-section view similar to that shown, multiply each square foot of area by 1.75 for actual surface area. Double for both sides. If the depth is 3" multiply by 1.5. Double for both sides.

CHAIN LINK FENCES

In estimating the paint requirements for chain link fences your first consideration should be the method of application. The most economical and recommended method is with an extra-long-nap roller. Measure length by height and multiply by 2 for both sides.

FIGURE 20.3 How to calculate amount of paint needed for corrugated surfaces and chain link fencing. *(continued)*

Outside Surface Area of Pipe

Outside Diameter	Square feet per lineal foot
0.5 inches	0.13
1	0.26
1.5	0.39
2	0.52
2.5	0.65
3	0.79
4	1.05
5	1.31
6	1.57
7	1.83
8	2.09
9	2.36
10	2.62

Metric Conversions

Linear Measurements

1 mil	=	25.4 microns
1 inch	=	2.54 centimeters
1 foot	=	0.3048 meters
1 yard	=	0.9144 meters
1 mile	=	1609.3 meters

Area measurements

1 square inch	=	6.452 square centimeters
1 square foot	=	929 square centimeters
1 square yard	=	0.8361 square meters
1 square mile	=	2.59 square kilometers

Liquid Measurements

1 quart	=	0.9463 liters
1 gallon	=	3.7853 liters

Weight Measurements

1 pound	=	119.8 grams

FIGURE 20.4 How to calculate amount of paint needed for pipe.

- To see what the color will actually look like when dry, apply paint with a brush on a white blotter; b the oils will quickly be absorbed, leaving the paint the color it will be when dry on the wall.

- When repainting wood baseboard, chances are there will be a build-up of wax from applications on adjacent hardwood or resilient floors; remove the wax with a cleaning solution and sand lightly for a good adhering quality.

- When using a dark finish color over a white primer, add some of the darker finish to the primer to tone it down a bit.

- When repainting exterior wood siding, don't sand with a disc sander before re-painting because the rotary motion will more than likely gouge the wood.

- Paint a chain link fence with a sponge; you'll save paint by avoiding the splatter that comes with using a brush.

Brushes and Rollers

Brushes come in many widths and two basic shapes, flat and angular. Angular brushes are good for trim work, and flat brushes are good for everything else. Large surfaces should be painted with a flat "wall" brush, a brush that is ⅞" (22.23mm) to 1" (2.54cm) thick and from 3 (7.6mm) to 6 inches (15.24 cm) wide.

Medium-size areas such as cabinets and furniture require a flat brush that is 2" (5.08cm) wide, and for small areas such a window or door trim a 1" (2.54cm)- or 2" (5.08cm) angular brush will do the trick.

Bristles can be either natural, the foremost being China white or black and Ox, or man-made, nylon or polyester.

Synthetic-bristle brushes are good for latex and water-based paints; natural-bristle brushes are good for alkyd and oil-based paints.

Brush Tips

- When buying a quality brush, examine its bristles and look for split ends, called "flags"; on premium brushes at least half the bristles are "flagged," which enables the brush to hold more paint and distribute it smoothly.

- A good paint brush will have springy bristles.

fastfacts

When you buy a new brush for oil paint, make it last longer by placing it in a can of linseed oil for 24 hours.

- Hit the brush on your hand to see if any bristles come out easily; if they do, buy a better brush.
- Angular brushes either 1" (2.54cm) or 2 " (5.08cm) wide are the best for wood trim work.

WALL COVERINGS

The first decorative wallpaper dates back to 15th century Europe and was either hand-painted or stenciled. Although the Chinese are often credited with inventing wallpaper, there is no evidence to indicate that its introduction in Asia predated European manufacture. By the 17th century, highly decorated wallpapers were imported from China; however they were referred to as India papers. At the end of the 18th century the first machine for the manufacture of wallpaper was introduced. At this time the French were known for their fine designs and the English for their production techniques. The 19th and 20th centuries brought mass production to the industry and advances in stain-resistant papers and in the manufacture of vinyl-coated and solid vinyl wall coverings.

fastfacts

After you have cleaned your brush, give it a rinse in a fabric-softener solution to keep it soft.

Terminology

The following terms are used when selecting wall coverings:

- Washable: A wall covering that can withstand occasional sponging with mild detergent
- Scrubbable: The wall covering can withstand light scrubbing with a brush and water with a mild detergent.
- Stain-resistant: The wall covering will show no appreciable change after stains such as grease, light oil, coffee, and most food are removed.
- Abrasion-resistant: The wall covering will stand certain abrasive actions such as rubbing, scrubbing, or light scraping.
- Colorfastness: The ability to retain color when exposed to light over a specific period of time.
- Peelable: The decorative surface of a wall covering may be dry-peeled from its substrate (backing); these coverings are generally paper-backed vinyl
- Strippable: The wall covering along with its substrate can be dry-peeled; usually fabric-backed vinyl.

Wallpapers can be classified into three types:

- Non-washable or water sensitive: Must be cleaned with a commercial cleaner
- Water resistant: Can usually be cleaned one to three times with a mild soap or detergent
- Scrubbable: Truly washable

fastfacts

13 square feet (1.21 square meters) = one lineal yard (.914 meters) of 54-inch-wide (1.37 meters) wall covering

9 square feet (.837 square meters) = 1 lineal yard (.914 meters) of 36-inch-wide (.91 meters) wide wall covering

Some wallpapers are treated with a plastic resin that improves strength, and others are treated to prevent mold and mildew. Strippable paper applied over an oil-base primer can be carefully pulled from the wall and reused.

Wallpapers are priced by the roll, and single rolls usually cover 30 (2.79 meters) to 35 square feet (3.25 square meters). A double roll covers 60 square feet (5.58 square meters) to 70 square feet (6.5 square meters).

Fabric wall coverings come with a paper back; grass cloth burlap, chintz, ticking, and canvas are examples. Felt wall coverings have acoustical qualities and are often treated to become soil- and flame-resistant.

Vinyl Wall Coverings

Vinyl wall coverings range from vinyl or Tedlar™-coated to solid vinyl with backing, all of which offer stain resistance. The vinyl layer in these wall coverings can vary from 2 to 35 mils in thickness. Vinyl wall coverings are sold in rolls of 52" (1.29 meters) to 54 inches (1.37 meters) wide and 30 feet (9.9 meters) in length.

There are three types of vinyl wall coverings:

- Type I: Lightweight, 7 to 13 ounces per square yard, designed for less heavily trafficked areas

- Type II: Medium weight, 13 to 22 ounces per square yard, designed for general use on walls where exposure to scuffing is expected

- Type III: Heavy duty, more than 22 ounces per square yard, designed for areas where rough abrasion will take place, such as use as wainscoting or in heavily trafficked corridors

Vinyl wall covering installation relies on good surface preparation. All walls must be dry, and free of grease, mildew and mold, and stains. When installed over gloss or semi-gloss paint, the painted areas must be abraded by sanding with a light sandpaper, followed by a coat of adhesion-promoting primer. Walls previously covered with wallpaper must be sanded after the old paper is removed or cleaned with an adhesive remover to prevent the growth of mold.

The choice of adhesive for vinyl wall coverings is important. These adhesives vary in their level of strippability, wet tack, and ease of use, so it is important to read the manufacturer's instructions for use.

FLOORING

There are many flooring options available to light commercial and residential contractors and homeowners. Cost of materials and installation labor runs the gamut from vinyl composition tile (VCT) to stone veneer set in mortar.

Resilient Flooring

This category encompasses vinyl composition tile, solid vinyl tile, sheet vinyl, and linoleum.

Vinyl Composition Tile (VCT)

This type of resilient flooring, a composition of thermoplastic binders (polyvinyl chloride resins), fillers, and pigment (for color), is one of the most cost-effective floor coverings. The size of the individual tiles are generally 12" x 12". Thickness availability is ⅛" (3.18mm), ³⁄₃₂" (2.38mm), and ¹⁄₁₆" (1.59mm). VCT is available as Type 1-smooth surface and Type 2-embossed surface.

Solid Vinyl Tile

A more expensive and more durable version of a standard 12" (30.48cm) x 12" (30.48cm) individual tile, solid vinyl tiles are available with a smooth or embossed surface. Class A solid vinyl flooring contains a constant composition throughout its thickness. Class B tile consists of layers of binders, fillers, and color with a vinyl resin

fastfacts

When installing resilient flooring over a new concrete slab, test for moisture in the slab by using a Delmhorst moisture meter. If there is a "red" reading, the concrete is wet; a green reading is O.K. but a subsurface reading should follow. Drill two 1/8" (3.175mm) holes in the slab, insert probes. A reading in the 85-95 range indicates low moisture levels between 2-4%.

content not less than 60% of the weight of the binder. Class C tiles can be manufactured to either Class A or Class B specifications but, additionally, have a permanently bonded protective coating overlay.

Depending upon the manufacturer, flooring tiles are also available in 16 inch (40.64 cm) or 18 inch (45.7 cm) squares.

Sheet Vinyl

Sheet vinyl flooring is classified by grade as to wear type, backing type, minimum wear layer thickness, and minimum overall thickness.

Grade	Wearlayer thickness	Overall thickness
A	.050" (1.270mm)	.080" (2.032mm)
	.020" (.508mm)	.060" (1.524 mm)
B	.030" (.762mm)	.060" (1.524mm)
	.014" (.356mm)	.050" (1.270mm)
C	.020" (.508mm)	.050" (1.270 mm)
	.010" (.254mm)	.050" (1.270mm)

Note: The different wearlayer thickness available in each of the three groups relates to whether the tiles are translucent, solid, or decorative with visability through the wear layer. Sheet vinyl goods are available in 6 ft (1.8 meters) and 12 ft (3.6 meter) wide rolls.

Linoleum

This type of flooring originated in Great Britain in the 1800s and was in widespread use in the U.S. throughout the first half of the 20th Century. It is no longer manufactured in America. Linoleum is

fastfacts

If the house is damp and has not been closed in, it is wise to check the moisture level of the subfloor. Using a Delmhorst Moisture meter's wood scale, a Green reading would indicate 6-15% Moisture (O.K.), a Yellow reading would indicate 15-17% moisture (allow to dry a little longer) and a Red reading would represent more than 17% moisture too much, needs considerable drying).

once again being offered to consumers, as an import from Europe, in roll form, approximately 6'6" wide (1.96 meters) and in squares measuring 23" (58 cm) x 23" (58cm) and 13" (33cm) x 13"(33cm).

Wood Flooring

Oak is the dominant species for wood flooring, although maple, chestnut, teak, exotic hardwoods and even pine are among other widely available choices.

Oak flooring has four basic grades:

- Clear oak - consists mainly of heartwood with a minimum of character marks or discoloration and exhibiting a uniform appearance, allowing for normal heartwood color variations. Small burls, fine pin worm holes are acceptable as well as small tight checks.

- Select oak - this grade of oak can contain slight milling imperfections, small tight knots every 3 feet (91.44cm), pin worm holes, and a reasonable amount of slightly open checks. Slight imperfection in face work (torn grain) permitted, a brown machine burn across the face not exceeding ¼ inch (6.35mm) width is allowed.

- No.1 Common oak - wood containing prominent variations in color and varying characteristics. Cannot have broken knots over ½ inch (1.25 cm). Minor imperfections in machining allowed as is an occasional dark machine burn not exceeding ½" (1.25cm) wide.

- No. 2 Common oak - Large broken knots, dark machine burning exceeding ³⁄₆₄" (2.38 mm) deep is allowed and may contain sound natural variations. This grade of oak finds application in general utility use.

fastfacts

Many builders apply only two coats of surface finish (wear coats) on new oak flooring. Specifying three coats will result in a superior finish with longer life.

fastfacts

"Stain resistant" is a chemical treatment offering protection from staining, while "soil resistance" is achieved by coating the surface of the carpet fibers to prevent soil particles from clinging to the fibers, thereby making vacuuming more effective

Carpet

With two basic types of materials, natural fibers and synthetic fibers, the simplicity ends there. Various types of weaves, patterns, and colors leave the consumer a wide range of quality, texture, and color options- and price.

Methods of carpet manufacture include:

- Tufted-a high-speed method by which the yarns are inserted through a prewoven backing fabric leaving the stitches long enough to be either cut off or left as loops woven in an in-and-out method of interlacing both surface and backing yarns in one operation.
- Knitted - the surface and backing loops are woven together with a stitching yarn on a machine with three sets of needles. As in weaving, this type of carpet is manufactured in one operation.
- Fusion bonded - two backing fabrics running parallel with a space in between are used and the backing has an adhesive on its face. Implanting a multifold fabric web between the backings creates a sandwich and when the blade slices through the middle of the sandwich, two identical sections of carpet are created.

Some of the textures created by carpet manufacturers are:

- Cut pile - made from unset yarns to create an even, velvety texture. Can also be created from firm ended yarns. Cut pile carpets look luxurious, but show footprints easily.
- Level loop pile - loops are all of the same height and are created by atufting, weaving, or knitting action. There is some variation in height of the loops and, while suitable for heavily trafficked areas, the space in between the loops tends to collect dirt.

- Cut and Loop - by creating different loop heights, a variety of textures can be created.

Seamless flooring is created by a process whereby a thick liquid-type material is spread over the flooring substrate, which when cured, creates flooring with no seams. This material is self-leveling and contains a resin matrix, either epoxy, one or two part polyurethane or polyester or one or two part neoprene (polychloroprene) material along with fillers and a decorative topping. This type of floor finds application in laboratories or where sanitary conditions are important (commercial kitchens/food storage areas) or where moisture or water could "lift" conventional resilient flooring with seams.

Ceramic tiles and stone veneer flooring come in a wide range of sizes, colors, textures, and costs. This makes any discussion of ceramic tile and stone veneer flooring far ranging. These types of flooring materials can be installed in one of two methods-thin-set or set in a mortar bed (Figure 20.5).

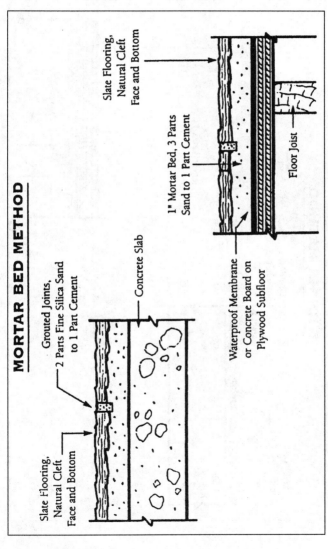

FIGURE 20.5 Thinset and mortar bed methods for tile/stone flooring. *(continued on next page)*

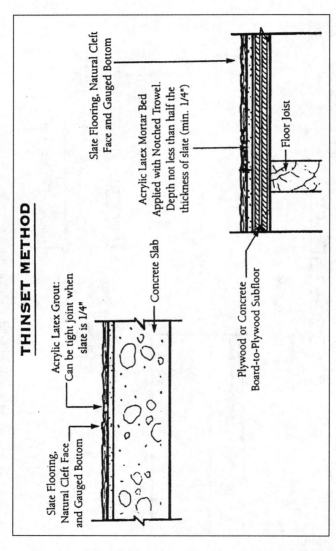

THINSET METHOD

Slate Flooring, Natural Cleft Face and Gauged Bottom

Acrylic Latex Grout: Can be tight joint when slate is 1/4"

Concrete Slab

Slate Flooring, Natural Cleft Face and Gauged Bottom

Slate Flooring, Natural Cleft Face and Gauged Bottom

Acrylic Latex Mortar Bed Applied with Notched Trowel. Depth not less than half the thickness of slate (min. 1/4")

Floor Joist

Plywood or Concrete Board-to-Plywood Subfloor

FIGURE 20.5 Thinset and mortar bed methods for tile/stone flooring. *(continued)*

21

WINDOWS, DOORS, HARDWARE

A discussion of windows ought to begin with the anatomy of a window, since terms such as "mutton" bars, double glazing, and Low E can be confusing. We all know what double- and single-hung windows are, and casements, awnings, and sliders are also familiar to most of us, as are the head, jamb and sill. But there is some confusion over other window parts and terminology.

- Air infiltration: The amount of air that passes between the window's sash and its frame
- Blind stop: That portion of a double-hung window located between the jamb and the sash, acting as a stop for the attachment of a storm sash
- Borrowed light: A fixed pane of glass installed in a frame in an interior partition, which allows light from one area or room to "borrow" light from the other
- Box size: The actual size of the window frame, excluding the exterior trim
- Brickmold: The molding around the window that is joined to the exterior window facing and acts as a finish boundary to the adjacent exterior wall
- Check rail: The bottom rail of the top sash and the top rail of the bottom sash, which meet when the window is closed. Also called a meeting rail

- Check stiles: The two vertical members of a sliding window that meet when the window is closed
- Cladding: The material that covers the exterior portion of a window; a wood window with an exterior face of aluminum is an aluminum-clad window
- Cottage style: Double-hung with the lower sash taller than the upper sash
- Direct set: Glass is set directly into the frame, forming a fixed glass window; the picture windows of the 1950 and '60s were direct set windows
- Double glazing: Not to be confused with insulating glass, this process involves placing a pane of glass in its own frame onto the back or front of a single pane window to create an airtight space for insulating purposes; this insert can be easily removed for cleaning, etc.
- Flankers: Windows set on either side of another window or door, such as windows on each side of a bay window
- Glazing: The act of installing glass in a window or frame
- Glazing bead: Also known as a glass stop, a long, thin piece of wood used to secure the glass in the individual window pane or sash
- Grills: Strips that divide the window into panes and snap into place, not to be confused with mullions or muntins (described below)
- High performance option: Generally refers to glazing options such as Low E, gas-filled, or tinted glass, all of which are energy conservation measures. Figure 21.1 contains heat-gain figures for high-performance and insulating glass as compared to single-pane glass)
- Insulating glass: Not to be confused with double glazing; two panes of glass, hermetically sealed, creating a dead-air, insulating space
- Laminated glass: two or more layers of glass with an inner layer of transparent plastic bonded together with heat and pressure, primarily, in the case of windows, to create a UV (ultraviolet) blocker
- Low E glass: Low E (emissivity) glass reduces heat transfer through the window; the lower the E rating, the lower the amount of radiated heat that will be allowed to pass through the window

Heat Gain and Performance Data

Heat Gain Data

In areas of the U.S. where cooling is the major energy cost, glazing may be the most important factor in energy-saving. That's because cooling costs are based almost solely on heat gains transmitted through the glass. The accompanying table is used to show maximum heat gain by type of glass.

Clear	Heat Gain	Tinted Grey/Bronze	Heat Gain	Medium Performance Reflective	Heat Gain
Single-pane ¾" or ⅞"	214	Single-pane grey ¾" (for comparison only)	165	Single-pane bronze (for comparison only)	106
Single-pane ¾" (for comparison only)	208	Single pane bronze ¾" (for comparison only)	168		
Double-pane (for comparison only)	186				
Double-pane high-performance insulating	113	Double-pane high-1 performance sun insulating			

FIGURE 21.1 Heat gain for various types of glazing.

- Masonry opening: Referred to simply as MO, this is the opening in a masonry wall large enough to accept a window or door frame, allowing extra space for alignment and a surface or blocking to secure the window; a rough opening is the opening in a wall other than a masonry wall, such as a wood or metal-framed wall.

- Mull casing: A wood or aluminum (if aluminum window) molding that covers the joint when two windows are placed side by side.

- Mullion: The vertical frame piece that joins separating windows, doors, or panels set in series.

- Muntin: A short bar that separates glass panes in a window.

- Nailing Fin: A strip of material protruding from each side of a window that is used as a nailing base to secure it in an opening.

- Operating panel: The panel in a sliding door that opens and closes.

WINDOW CONFIGURATIONS

From simple single-hung to elaborate custom-designed fenestration (the fancy name for the arrangement and design of windows in a building), window configurations are nearly limitless. The more common types are:

- Single- or double-hung: sometimes referred to as vertical sliders; the single-hung type has only one movable sash, while the double-hung has both upper and lower movable sashes.

- Horizontal sliders: contains a manually operated sash that slides horizontally.

- Casement type: one or two sashes hinged on the window-frame jamb and opened outward, either by pushing them open or turning an operator.

- Awning type: just like an awning that projects out from a horizontal frame, some awning windows project outward from the top just like a canvas awning, while others project out from a bottom horizontal member, forming a scoop that needs to be closed during a rain storm.

- Fixed window: also known as a "fixed light," it is just a pane of glass fixed in a frame that is inoperable.

fastfacts

The symbol "X" on a sliding window or door signifies the movable sash or portion, while an "O" designation is fixed. A set of patio doors labeled XOX would designate a center fixed panel and movable panels on each side.

STANDARDS FOR WOOD WINDOWS

The National Fenestration Council has various window ratings based upon air infiltration, water infiltration, and physical load:

Grade 20: Suitable for residential construction

- Air infiltration: At wind pressure of 25 mph, not more than .34 cubic feet per minute (CFM) per lineal feet of sash crack perimeter is allowed.

- Water infiltration: No water shall pass beyond the interior of the unit in a 15-minute test with 5 gallons per hour per square foot (equals to 8 inches of rain per hour) under pressure of 34 mph.

- Physical load: 89 mph wind placed against the window, held for 10 seconds, and released; no glass breakage, hardware damage, or deformation shall result in malfunction.

fastfacts

When you see a window diagram with a dotted triangle in it, it indicates how the window opens. The point of the triangle indicates the hinge side.

 An awning window opens outward from top.

 A casement window on opens out toward the center pane.

Grade 40: For use in light commercial construction

- Air infiltration: Same as above except not more than .25 cubic feet per minute.
- Water infiltration: Same as above under air pressure of 42 mph.
- Physical load: Same as above except testing done under air pressure of 126.5 mph.

Grade 60: For heavy commercial construction

- Air infiltration: Same as above but no more than .10 cubic feet per minute.
- Water infiltration: Same as above except under air pressure of 50 mph.
- Physical load: Same as above except testing done under air pressure of 154.9 mph.

THERMAL MOVEMENT, THERMAL BREAKAGE, AND GLASS DEFLECTION

Have you ever seen a crack in one corner of your window and wondered how it got there?

Thermal expansion, developed either from internal stress or actual expansion of the glass, may be the answer. Thermal breakage can occur when the center of the pane of glass, heated by the sun, becomes much hotter than the edges. The center of the glass wants to expand, but it is restricted by the cool edges and by being fitted too tightly into its frame (Figure 21.2). The resulting stress can cause the glass to crack.

Deflection of the glass can also be affected by wind loads, both positive and negative (Figure 21.3), and, lastly, deflection from loads imposed on the frame (Figure 21.4) can also be responsible for breakage.

DOORS

Wood, metal, fiberglass, solid-core, stave-core, mineral-core—the construction and style of interior and exterior doors is not a simple matter.

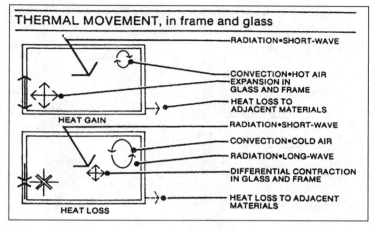

FIGURE 21.2 Thermal movement in glass in a frame.

Door construction involves:

- Materials of construction: wood, steel, fiberglass, medium-density fiberboard.
- Face material: the back and front panels visible to the eye—wood, metal, fiberglass, medium-density particleboard.

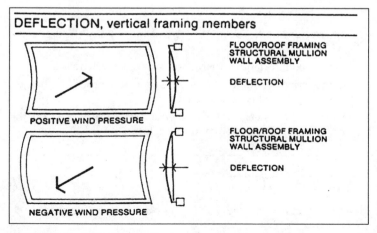

FIGURE 21.3 Deflection in glass in a frame due to wind load.

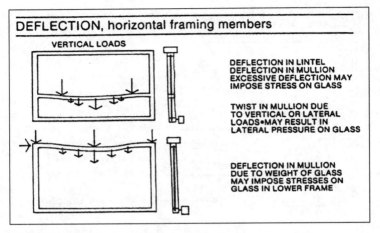

FIGURE 21.4   Deflection in a frame due to loading.

- Core material: the inner construction that provides rigidity and fire resistance when required—stave core (blocks of solid wood), particleboard, mineral board (fire-resistant).
- Architectural and aesthetic considerations: style, glass inserts, exotic veneers.

Doors, whether they are metal (referred to as hollow metal), wood, or fiberglass, all have the same basic components:

- Top, bottom, and side rails that provide rigidity (Figure 21.5a and 21.5b).
- Cores: Medium-density fiberboard, wood, particleboard.
- Internal reinforcement (Figure 21.6) for attachment of hinges, hardware, door closers, kick plates, etc.
- Faces: Wood veneers, medium-density fiberboard, steel (in the case of hollow metal doors), or fiberglass.

fastfacts

Glass with clean-cut edges has the greatest resistance to thermal breakage.

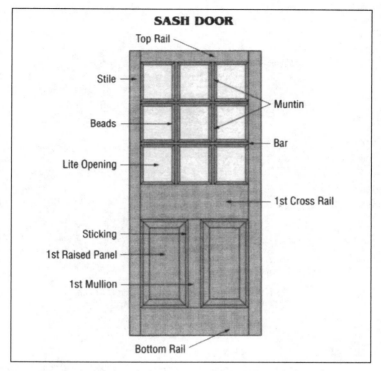

FIGURE 21.5A Basic components of a door with glass lites (sash).

Wood Doors

Wood door faces are often constructed of veneers such as birch, oak, mahogany, or even more exotic woods such as rosewood, cherry, or teak. Plain-faced wood doors, called flush doors, generally have a combination of wood, particleboard, or MDF cores and medium-density fiberboard faces that provide a smooth surface for painting. Figure 21.7 reveals the construction of two types of flush doors, veneer-faced stave core and veneer-faced composite (particleboard). Many flush doors, particularly those with MDF faces, have a lightweight composite honeycomb core.

Storing, Installing, and Maintaining Wood Doors

Store doors on a flat surface in a dry, well-ventilated area. Doors should be kept at least 3½ inches off the floor and should

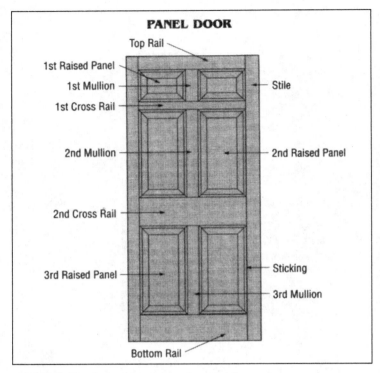

FIGURE 21.5B Basic components of a panel door.

have protective coverings under the bottom of the door and over the top. The covering should protect from dirt, water, and abuse, but should be loose enough to allow for air circulation.

Avoid exposure of interior doors to sunlight during storage; certain doors such as cherry, mahogany, walnut, and teak can discolor in their unfinished state when exposed to sunlight.

During installation be careful not to rack the doors when handling and fitting.

In fitting for height do not trim the top or bottom edge by more than ¾ inch unless additional blocking in these areas has been incorporated into the door design.

Drill pilot holes for all door attachments, hinges, closers, lockset/passage sets, kickplates, etc.

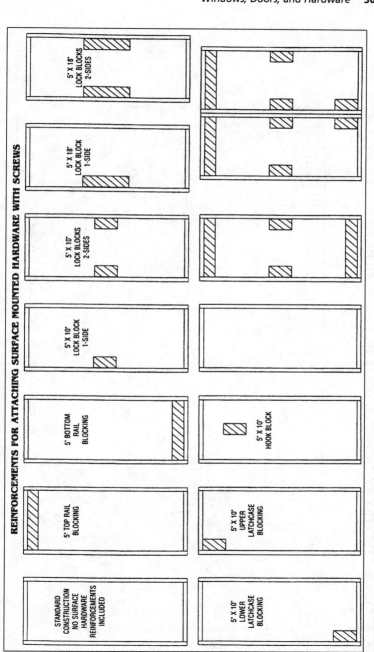

FIGURE 21.6 Hardware and special reinforcing requirements.

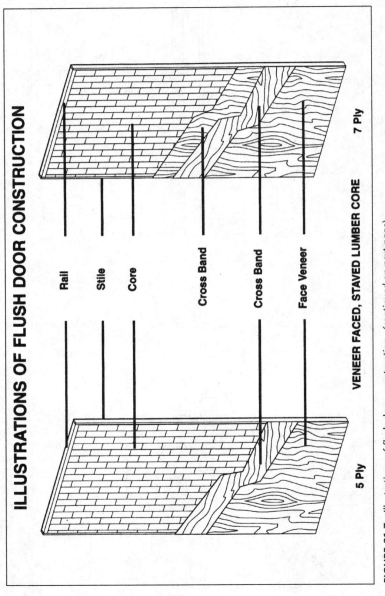

ILLUSTRATIONS OF FLUSH DOOR CONSTRUCTION

Rail

Stile

Core

Cross Band

Cross Band

Face Veneer

7 Ply

5 Ply

VENEER FACED, STAVED LUMBER CORE

FIGURE 21.7 Illustrations of flush door construction. *(continued on next page)*

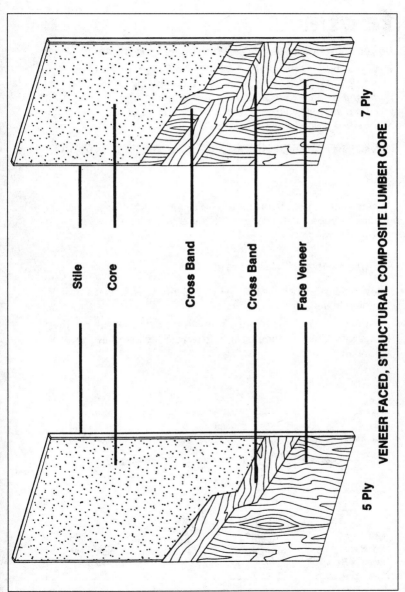

FIGURE 21.7 Illustrations of flush door construction. *(continued)*

fastfacts

Do you know how to determine which way a door "swings'?

With your back to the jamb where it will be hinged, if the door swings left, it is a left-hand swing; if the door swing right, it is a righthand swing. Figure 21.8 illustrates left or right handing when a door opens outward and reverse bevel handing, left and right (when a door opens inward).

When applying stain, or paint to the door, don't forget the top and bottom edges. Left unsealed to the environment, swelling and deterioration will occur.

Adjustment and Maintenance

- Allow at least ³⁄₁₆ inch clearance for swelling when a wood door in installed in very dry conditions.

- Ensure that all doors swing freely and do not bind on their hinges; if a door is hinge bound, it can be corrected as shown in Figure 21.9

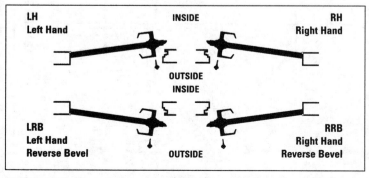

FIGURE 21.8 Door "handing", left, right, left hand reverse, right hand reverse.

Hinge Binding against Rabbet

Normally, hinge bind is found between the door and rabbet. There are several ways of shimming which will move the door in different directions. The following guidelines should be used in shim applications.

1. A shim can be placed between the frame hinge reinforcement and the hinge leaf. This will move the door toward the strike jamb. However, the hinge notch face gap will be increased and the hinge leaf surface will not be flush with the rabbet surface.

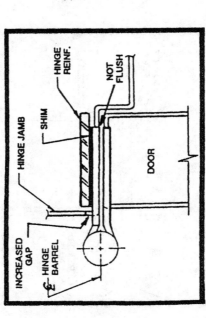

FIGURE 21.9 Correcting a hinge bound door. *(continued on next two pages)*

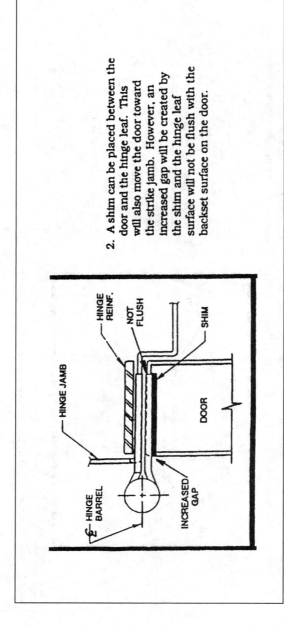

2. A shim can be placed between the door and the hinge leaf. This will also move the door toward the strike jamb. However, an increased gap will be created by the shim and the hinge leaf surface will not be flush with the backset surface on the door.

FIGURE 21.9 Correcting a hinge bound door. *(continued)*

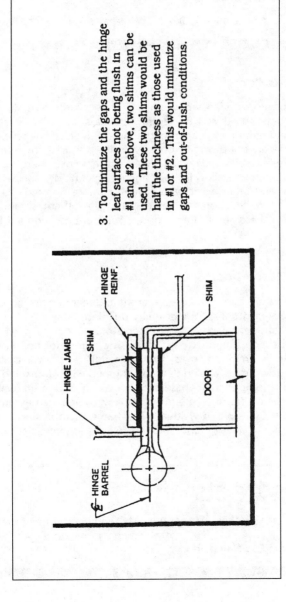

3. To minimize the gaps and the hinge leaf surfaces not being flush in #1 and #2 above, two shims can be used. These two shims would be half the thickness as those used in #1 or #2. This would minimize gaps and out-of-flush conditions.

FIGURE 21.9 Correcting a hinge bound door. *(continued)*

fastfacts

Hang the door before finishing it. Then remove to finish properly.

Fiberglass Doors

The art of reproducing wood-grain finishes is clearly visible in the paneled doors produced by several fiberglass door manufacturers. Only a very minute inspection and a tap on the face will reveal that these doors are not painted wood. Typically, polyurethane is foamed between wood stiles and rails to provide the core of a fiberglass door. The wood members allow for solid anchoring of hinges, locksets/passage sets, and other accessories. Tough compression-molded faces are bonded to these cores to create a strong, weather-resistant product, and in the case of interior doors, color often is bonded in the face slabs.

Hollow Metal Doors

These doors are frequently used as front and rear exterior doors and as entry doors into a garage or utility building. They are often used in light commercial construction for storage rooms, mechanical equipment rooms, and exterior doors. The frames for hollow metal doors are also referred to as "hollow metal" and are available as knocked-down parts (called KD) and shipped with jambs and head as separate components. These parts are assembled in the wall and secured to the framing members. Hollow metal frames are also available as "set-up-and-welded." The frame members are welded together at the factory and shipped fully assembled. Set-up-and-

fastfacts

Before applying that first coat of finish, sand the surface of the door lightly with 5/0 (180 grit) sandpaper to remove finger-prints and handling marks.

fastfacts

When finishing a door with a glazed section, flow the finish coat ever so slightly onto the glass. This will protect the putty and further waterproof this joint.

welded frames are generally used where the door and frame will get frequent and heavy usage and for exterior doors; KD frames are usually ordered for interior doors. Hollow metal doors and frames are available in a variety of metal gauges ranging from light-duty—22/24-gauge to heavy-duty—16/18/20-gauge—and in many widths and heights (Figure 21.10).

Hardware

Hinges, lock sets, and passage sets are available in a number of finishes, ranging from brass plate to solid brass, bronze, chrome, aluminum, and stainless steel. Architectural hardware finishes are designated by US numbers:

Bright brass, clear-coated	US3
Satin brass, clear-coated	US4
Bright bronze, clear-coated	US9
Satin bronze, clear-coated	US10
Satin bronze,dark oxidized, oil rubbed	US 10B
Bright chrome, plated	US26
Satin aluminum, clear anodized	US27
Satin chrome, plated	US26D

fastfacts

Did you know that your warranty will be void if you have not coated all six sides (front, back, four edges) of your door?

Standard Opening Sizes for Hollow Metal Doors

STANDARD OPENING SIZE

Opening Widths	Opening Heights 1 3/4 " Doors				1 3/8 " Doors	
2'0"	6'8"	7'0"	7'2"	7'10"	8'0"	8'10"	10'0"	6'8"	7'0"
2'4"	6'8"	7'0"	7'2"	7'10"	8'0"	8'10"	10'0"	6'8"	7'0"
2'6"	6'8"	7'0"	7'2"	7'10"	8'0"	8'10"	10'0"	6'8"	7'0"
2'8"	6'8"	7'0"	7'2"	7'10"	8'0"	8'10"	10'0"	6'8"	7'0"
2'10"	6'8"	7'0"	7'2"	7'10"	8'0"	8'10"	10'0"	6'8"	7'0"
3'0"	6'8"	7'0"	7'2"	7'10"	8'0"	8'10"	10'0"	6'8"	7'0"
3'4"	6'8"	7'0"	7'2"	7'10"	8'0"	8'10"	10'0"		
3'6"	6'8"	7'0"	7'2"	7'10"	8'0"	8'10"	10'0"		
3'8"	6'8"	7'0"	7'2"	7'10"	8'0"	8'10"	10'0"		
3'10"	6'8"	7'0"	7'2"	7'10"	8'0"	8'10"	10'0"		
4'0"	6'8"	7'0"	7'2"	7'10"	8'0"	8'10"	10'0"		

FIGURE 21.10 Standard size hollow metal doors.

fastfacts

Storm doors ought to be vented to eliminate excessive heat build-up when exposed to the sun.

Bright stainless steel	US32
Stan stainless steel	US32D
Powder-coated aluminum	MAL
Powder-coated bronze	M13

Passage and Lock Sets

A passage set is ostensibly the same as a lock set except that it has no keyed locking capability. Passage sets, however, may have a center push or turn button that allows it to be locked from one side only; this is called a privacy set and is installed on bathroom and some bedroom doors.

Locksets can be either cylindrical with a round knob; this also has a deadbolt), a tulip-shaped knob(Figure 21.11), a lever handle (Figure 21.12), or a mortise type (Figure 21.13) ; this one has a lever handle), but they all share much the same basic components, as can be seen in these exploded views.

A dead-bolt throw (Figure 21.14) is often used in conjunction with exterior door cylindrical locks, giving the homeowner an added sense of security.

Lock and passage sets can be ordered with a wide variety of knobs and handles (Figure 21.15); each manufacturer has a slightly different twist on design.

Lock and passage sets are primarily identified by function—how they are expected to work. Will one side be key-operated while the other has a thumb turn button; will the outside be key-lockable but the inside always open (so if someone inadvertently finds him- or herself locked in, he or she can exit by merely turning the knob)? Figure 21.16) contains a list of some of these non-keyed and keyed functions.

(text continues on page 391)

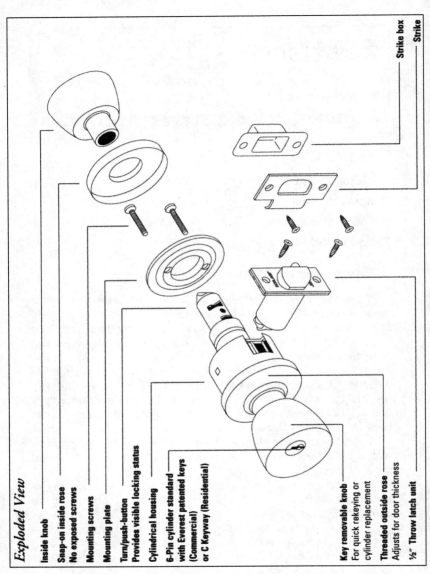

Exploded View

- Inside knob
- Snap-on inside rose
  No exposed screws
- Mounting screws
- Mounting plate
- Turn/push-button
  Provides visible locking status
- Cylindrical housing
- 6-Pin cylinder standard
  with Everest patented keys
  (Commercial)
  or C Keyway (Residential)
- Key removable knob
  For quick rekeying or
  cylinder replacement
- Threaded outside rose
  Adjusts for door thickness
- ½" Throw latch unit

- Strike box
- Strike

FIGURE 21.11 Lockset with a tulip shaped knob. *(By permission-Schlage division of the Ingersoll Rand Corp.)*

FIGURE 21.12 Lockset with a lever handle. *(By permission-Schlage division of the Ingersoll Rand Corp.)*

Exploded View

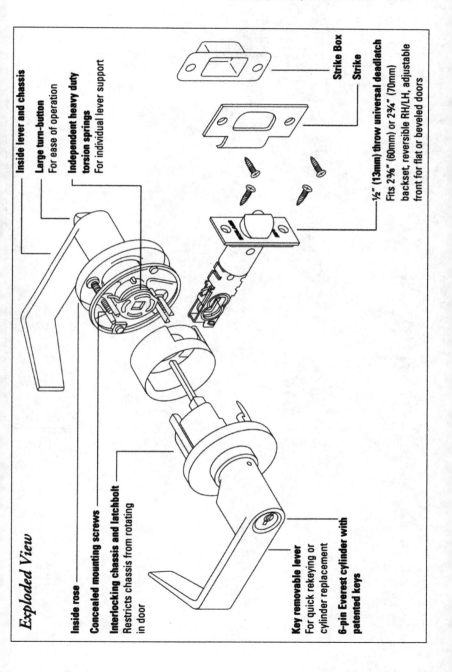

Inside lever and chassis

Large turn-button
For ease of operation

Independent heavy duty torsion springs
For individual lever support

Strike Box

Strike

½" (13mm) throw universal deadlatch
Fits 2⅜" (60mm) or 2¾" (70mm) backset, reversible RH/LH, adjustable front for flat or beveled doors

Inside rose

Concealed mounting screws

Interlocking chassis and latchbolt
Restricts chassis from rotating in door

Key removable lever
For quick rekeying or cylinder replacement

6-pin Everest cylinder with patented keys

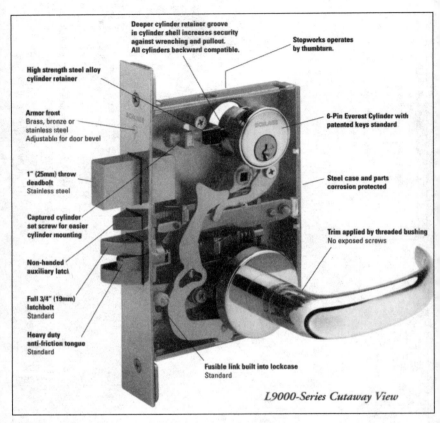

Deeper cylinder retainer groove in cylinder shell increases security against wrenching and pullout. All cylinders backward compatible.

Stopworks operates by thumbturn.

High strength steel alloy cylinder retainer

Armor front
Brass, bronze or stainless steel
Adjustable for door bevel

6-Pin Everest Cylinder with patented keys standard

1" (25mm) throw deadbolt
Stainless steel

Steel case and parts corrosion protected

Captured cylinder set screw for easier cylinder mounting

Non-handed auxiliary latch

Trim applied by threaded bushing
No exposed screws

Full 3/4" (19mm) latchbolt
Standard

Heavy duty anti-friction tongue
Standard

Fusible link built into lockcase
Standard

L9000-Series Cutaway View

FIGURE 21.13 Mortise lock with lever handle. *(By permission-Schlage division of the Ingersoll Rand Corp.)*

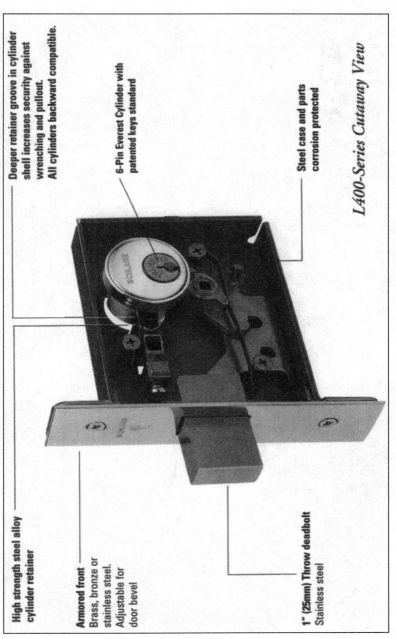

High strength steel alloy cylinder retainer

Deeper retainer groove in cylinder shell increases security against wrenching and pullout. All cylinders backward compatible.

6-Pin Everest Cylinder with patented keys standard

Steel case and parts corrosion protected

Armored front
Brass, bronze or stainless steel. Adjustable for door bevel

1" (25mm) Throw deadbolt
Stainless steel

L400-Series Cutaway View

FIGURE 21.14 A dead bolt throw. *(By permission-Schlage division of the Ingersoll Rand Corp.)*

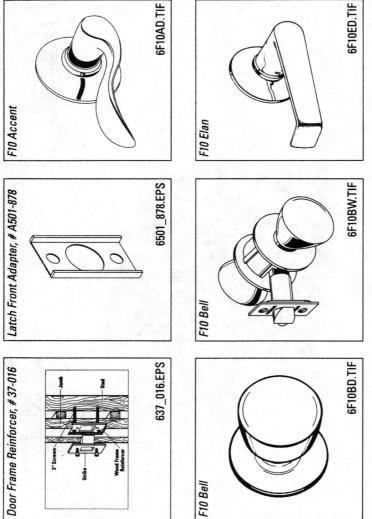

FIGURE 21.15 A wide variety of knob styles available for locksets and passagesets *(By permission–Schlage division of the Ingersoll Rand Corp., continued on next six pages).*

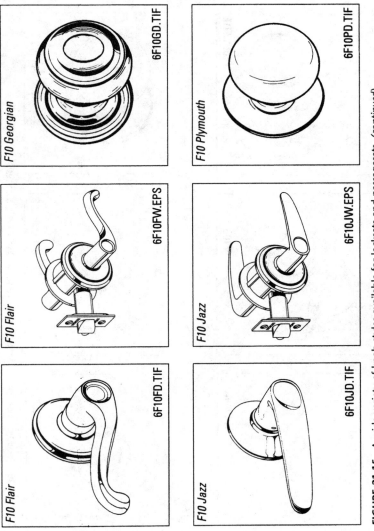

FIGURE 21.15 A wide variety of knob styles available for locksets and passagesets. *(continued)*

FIGURE 21.15 A wide variety of knob styles available for locksets and passagesets. *(continued)*

F40 Accent

6F40AW.TIF

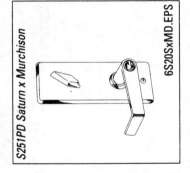

S251PD Saturn x Murchison

6S20SxMD.EPS

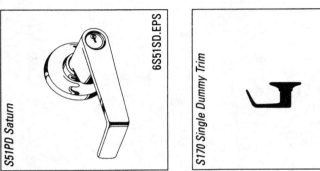

S51PD Saturn

6S51SD.EPS

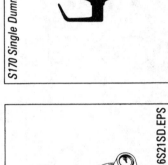

F40 Accent

6F40AD.EPS

S251PD Saturn

6S21SD.EPS

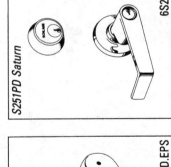

S170 Single Dummy Trim

7S170.EPS

FIGURE 21.15 A wide variety of knob styles available for locksets and passagesets. *(continued)*

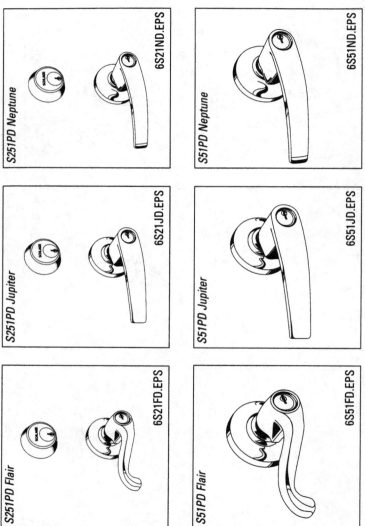

FIGURE 21.15 A wide variety of knob styles available for locksets and passagesets. *(continued)*

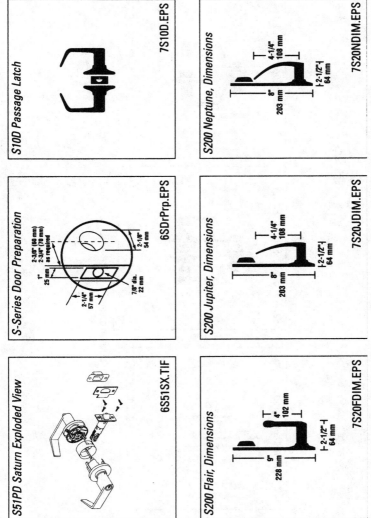

FIGURE 21.15 A wide variety of knob styles available for locksets and passagesets. *(continued)*

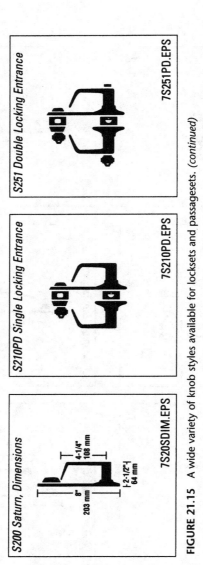

FIGURE 21.15 A wide variety of knob styles available for locksets and passagesets. (continued)

FIGURE 21.16 Choices of keyed and non-keyed lockset/passageset functions. (By permission-Schlage division of the Ingersoll Rand Corp.)

Functions

ANSI A156.2 Series 4000 Grade 2

Non-Keyed Functions

SCHLAGE   ANSI

A10S   F75

Passage Latch
Both knobs always unlocked.

A25D

Exit Lock
Blank plate outside. Inside knob always unlocked. Specify door thickness, 1⅜" or 1¾".

A30D   F77

Patio Lock
Push-button locking. Turning inside knob or closing door releases button, preventing lock-out.

A40S   F76

Bath/Bedroom Privacy Lock
Push-button locking. Can be opened from outside with small screwdriver. Turning inside knob or closing door releases button.

A43D   F79

Communicating Lock
Turn-button in outer knob locks and unlocks knob and inside thumbturn.

A170

Single Dummy Trim
Dummy trim for one side of door. Used for door pull or as matching inactive trim.

Keyed Functions

SCHLAGE   ANSI

A53PD   F109   **Entrance Lock**

Turn/push-button locking: pushing and turning button locks outside knob requiring use of key until button is manually unlocked. Push-button locking: pushing button locks outside knob until unlocked by key or by turning inside knob.

A70PD   F84   **Classroom Lock**

Outside knob locked and unlocked by key. Inside knob always unlocked.

A79PD   **Communicating Lock**

Locked or unlocked by key from outside. Blank plate inside.

A80PD   F86   **Storeroom Lock**

Outside knob fixed. Entrance by key only. Inside knob always unlocked.

A85PD   F93   **Hotel/Motel Lock**

Outside knob fixed. Entrance by key only. Push-button in inside knob activates visual occupancy indicator, allowing only emergency masterkey to operate. Rotation of inside spanner-button provides lock-out feature by keeping indicator thrown.

FIGURE 21.16 Choices of keyed and non-keyed lockset/passageset functions. *(continued)*

Hinges

Selecting the proper type, size, and number of hinges is very important in order to ensure long, trouble-free operation of any swinging door. Location and number of hinges are to be considered. Figure 21.17 illustrates the proper location for two or three hinges, guidelines for the number of architectural hinges, and construction materials for various functions.

GUIDELINES FOR NUMBER OF ARCHITECTURAL HINGES
1. Doors up to 60"...... 2 hinges
2. Doors 60" to 90"...... 3 hinges
3. Doors 90" to 120" .... 4 hinges

LOCATION OF ARCHITECTURAL HINGES

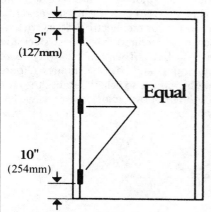

5"
(127mm)

10"
(254mm)

Equal

Top hinge 5" from jamb rabbet to top of barrel

Bottom hinge 10" from bottom edge of barrel to finished floor

Third hinge centered between top and bottom hinges

Note: Certain western states use a standard 7" from top and 11" From the bottom

GUIDELINES FOR ARCHITECTURAL HINGE METAL

1. Interior doors or non-corrosive areas use: Plated or painted Steel
2. Interior labeled door use: Plated or Painted Steel, Stainless Steel
3. Interior doors in corrosive areas use: Stainless Steel , Brass, Bronze
4. Exterior doors use: Stainless Steel, Brass, Bronze

FIGURE 21.17 Location of hinges and guidelines for number of hinges depending upon door size.

fastfacts

Full-mortise, knuckle-type hinges are referred to as "butts."

- Two hinges for solid core doors up to 60 inches in height.
- Three hinges for solid core doors up to 90 inches in height.
- An additional hinge for every additional 30 inches in height.
- Interior hollow core doors weighing less than 50 pounds and not over 90 inches in height can be hung with two hinges.
- Heavyweight doors over 175 pounds require heavyweight hinges.
- For special doors, consult the door manufacturer for hinge size and number.

Mortise-type hinges, or butts, are those that are recessed into both door jamb and door (Figure 21.18) and are most frequently used on exterior doors or heavy interior architectural doors. Heavy-duty mortise hinges are often manufactured in stainless steel and can be purchased as ball bearing butts. There are many types of hinges; spring hinges, pivot hinges, and invisible hinges (Figure 21.19). And then there are hobby hinges, so called because they are decorative and fit the needs of the hobby cabinet maker as well as the professional millwork shop.

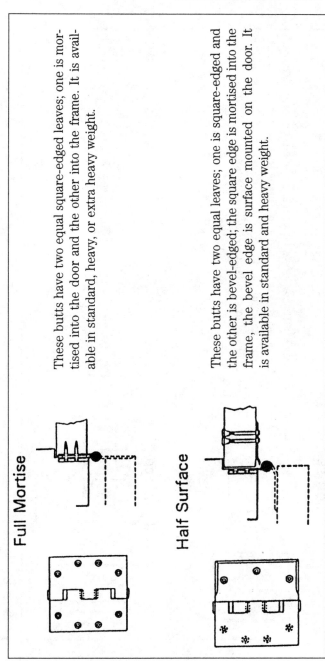

Full Mortise

These butts have two equal square-edged leaves; one is mortised into the door and the other into the frame. It is available in standard, heavy, or extra heavy weight.

Half Surface

These butts have two equal leaves; one is square-edged and the other is bevel-edged; the square edge is mortised into the frame, the bevel edge is surface mounted on the door. It is available in standard and heavy weight.

FIGURE 21.18 Mortise type hinges, also called butts. *(continued on next page)*

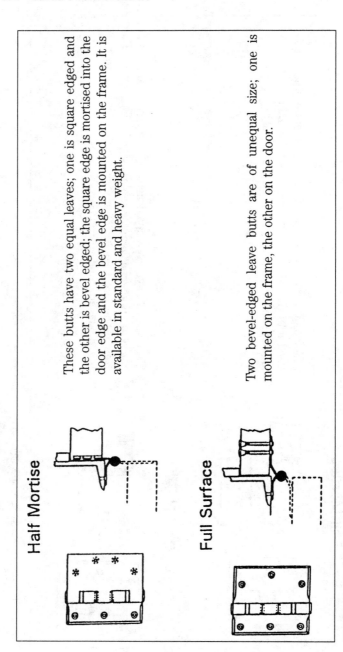

Half Mortise

These butts have two equal leaves; one is square edged and the other is bevel edged; the square edge is mortised into the door edge and the bevel edge is mounted on the frame. It is available in standard and heavy weight.

Full Surface

Two bevel-edged leave butts are of unequal size; one is mounted on the frame, the other on the door.

FIGURE 21.18 Mortise type hinges, also called butts. *(continued)*

Swing clear/full mortise are also available in half-surface, half-mortise, and full-surface configurations. These types of butts provide an unobstructed clear frame opening when door is in the 90° open position. It is available in either a single- or double-acting configuration, usually mortised into the door and frame, providing closing action without a separate closer.

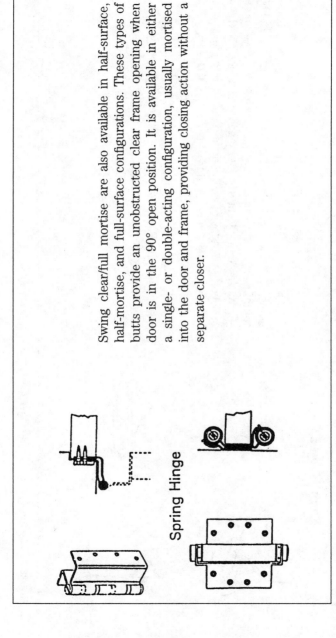

FIGURE 21.19 Spring hinges, pivot hinges, invisible hinges. *(continued on ext two pages)*

Offset pivot hinges are mortised into the top and bottom edges of the door and into the frame jamb at the top and bottom. These hinges can also be mortised into the floor and the top of the frame. Center pivot hinges are attached to the top and bottom edges of the door and either into the top and bottom of the frame or into the floor and the top of the frame. Fully mortised into the edge of the door and frame, the hinge portion is not visible when the door is closed, except when the Paumelle or Olive Knuckle hinge is used, the olive-shaped portion is visible as an architectural feature.

Pivot Hinges

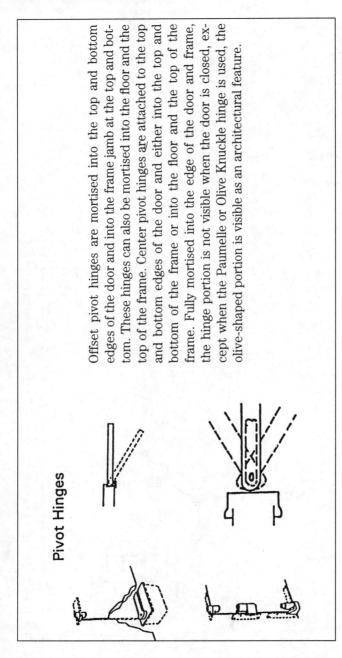

FIGURE 21.19 Spring hinges, pivot hinges, invisible hinges. *(continued)*

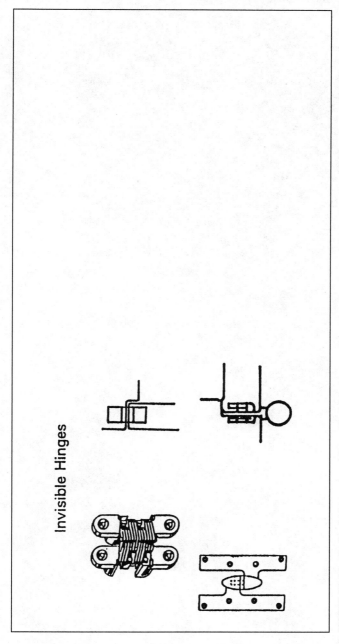

FIGURE 21.19   Spring hinges, pivot hinges, invisible hinges. *(continued)*

22

MILLWORK

The dictionary defines millwork as follows:

Products such as moldings, door frames, stair work and cabinets made in a woodworking plant.

It also includes shelving, laminated counter tops, custom-made doors, and windows.

Millwork involves not only the design and manufacture of these items, but an intimate knowledge of joinery and the characteristics and nature of a wide variety of softwoods and hardwoods. Familiarity with millwork begins with knowing a little about the make-up of the tree from which the wood will be milled.

WHAT IS A TREE?

The many parts of a tree, whether it be softwood or hardwood, present several distinctive appearances, depending upon not only the type of wood being harvested, but the way in which the tree is sawn at the mill. By cutting the log one way, a vertical grain board will emerge. Cutting the log another way, flat grain will present a different look when stained. Figure 22.1 depicts the various parts of a tree as it enters the mill.

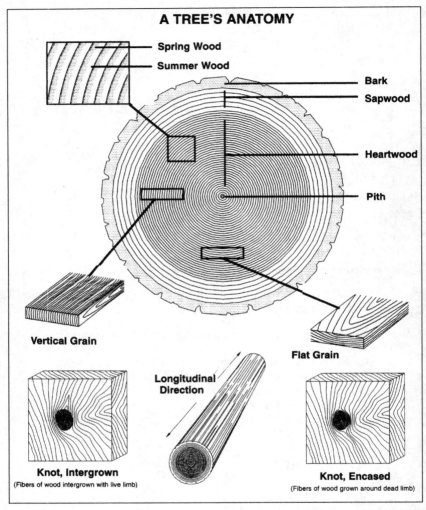

FIGURE 22.1 Various parts of the trees as it enters the mill.

When veneers are cut from a log, they are done so by peeling off thin layers of wood and, depending upon how the layers are peeled off, various visual effects of the veneer are created. Figure 22.2 illustrates several methods whereby logs are peeled to create rotary, plain-sliced, and quarter-cut veneers. Figure 22.3 shows the end result of each type of cut.

TYPES OF VENEER CUTS

The manner in which veneers are cut is an important factor in producing the various visual effects obtained. Two woods of the same species, but with their veneers cut differently, will have entirely different visual character even though their color values are similar.

In plywood manufacture, the principal methods of cutting veneers are used, depending on the type of veneer required (whether for face, crossband, or core), the nature of the log, and the veneer figure desired. Primarily the veneer slicer and veneer lathe are the equipment employed. The methods are:

ROTARY

The log is mounted centrally in the lathe and turned against a razor sharp blade, like unwinding a roll of paper. Since this cut follows the log's annular growth rings, a bold variegated grain marking is produced. Rotary cut veneer is exceptionally wide.

PLAIN-SLICING (OR FLAT-SLICING)

The half log is mounted with the heart side float against the guide plate of the slicer and the slicing is done parallel to a line through the center of the log. This produces a variegated figure.

FIGURE 22.2 Rotary, plain sliced, quarter sliced, rift cut veneers. *(By permission-Woodwork Institute, West Sacramento, Ca.)*

FIGURE 22.3 On right-the end result of each type of veneer cut. *(By permission-Woodwork Institute, West Sacramento, Ca., continued on next page)*

QUARTER-SLICING

The quarter log is mounted on the guide plate so that the growth rings of the log strike at approximately right angles, producing a series of stripes, straight in some woods varied in others. No limitations on the amount of medullary ray in oak.

RIFT-CUT HARDWOOD OR VERTICAL GRAIN CUT SOFTWOOD

RIFT-CUT veneer is produced in the various species of Oak. Oak has medullary ray cells which radiate from the center of the log like the curved spokes of a wheel. The rift or comb grain effect is obtained by cutting perpendicularly to these medullary rays either on the lathe or slicer. Comb grain is selected from rift. Medullary ray flake is limited.

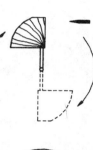

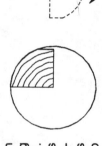

FIGURE 22.2 Rotary, plain sliced, quarter sliced, rift cut veneers.

FIGURE 22.3 On right—the end result of each type of veneer cut. *(continued)*

fastfacts

Heartwood is the inner layer of wood in growing trees that has ceased to contain living cells.

As the veneering process takes place, the thin layers of wood produced are retained in the sequence by which they were cut or peeled from the log. The yield from a single log is referred to as a "flitch" and all veneers of the particular flitch are stored in sequence so that if various matched grain patterns are required for wall paneling or furniture, they can be re-created by the assemblage of specific slices of wood.

THE HARDWOODS

The term *hardwood* does not denote hardness of wood, the term applies to wood harvested from deciduous, broad leaf trees, *Angiosperms,* if you want to get technical. Hardness of a piece of wood is defined as *resistance to indentation,* measured by the use of a modified Janka hardness test.

Confining ourselves to American hardwoods, there are 21 species of note used in the manufacture of millwork items. Some of the terms used to describe the characteristics of the following species may require an explanation:

Figure - pattern produced in wood by annual ring growth, knots, wavy- irregular, coloration, all of which are termed "figures".

Grain - when referred to as *straight,* means that fibers and other longitudinal configurations in the wood run parallel to the axis of the piece of lumber.

Movement - the potential for twisting, warping, or cupping.

Sapwood - the outer zone of wood in a tree, next to the bark.

Texture - relative size and distribution of wood elements in the log or board, such as *coarse* (large), *fine* (small) or *even* (uniform).

The Twenty One American Hardwoods-
Both Common and Botanical Names

Red Alder - *(Alnus rubra)*
A Pacific Northwest wood, almost white in color when sawn, changing to a light brown with yellow or reddish tinge. A straight-grained wood with uniform texture, this wood machines well and is excellent for turnings.

Ash - *(Fraxinus spp)*
An Eastern U.S. tree, another straight-grained wood that varies in color from nearly white to grayish or light brown heartwood. Ash machines well and dries easily with some movement in performance. Ash has good overall strength in relation to its weight and excellent shock resistance which makes it applicable for flooring, tool handles, and baseball bats.

Aspen - *(Populus tremuloides)*
Aspen, a straight-grained white-colored wood (heartwood is light brown), has a uniform texture. Aspen does not split when nailed, but it develops a slightly fuzzy texture when machined. This wood finds use in drawer sides, moldings, picture frames, and matchsticks. Because of its low conductivity of heat, aspen is often used in saunas.

Basswood - *(Tilla Americana)*
This wood is also known as Linden and is found principally in Northern and Lake states. Basswood machines well and is easy to work with, especially with hand tools, making it a favorite of carvers. It has fairly high shrinkage but good dimensional stability once it has dried.

Beech - *(Fagus gradifolia)*
Found in the Central and Mid-Atlantic region, beech is straight-grained with a close uniform texture, white in color with a red tinge. This wood is readily worked by machine or hand, but it does have a tendency to warp, split, and surface check. Beech, while used for furniture, millwork, and paneling, finds particular use in food containers since it imparts no odor or taste to the contents

Yellow beech - *(Betula alleghaniensis)*
Another principally Eastern tree, straight-grained with uniform texture, yellow beech takes stain extremely well. Yellow beech is a heavy, hard wood that dries slowly, has high shrinkage and so it is susceptible to movement in place. Kitchen cabinets, toys, turnings, and furniture are some of the products made from yellow beech.

Cherry - *(Prunus serotina)*

The main commercial growing areas for cherry are Pennsylvania, Virginia, West Virginia, and New York. The sapwood from cherry trees is creamy white, while the heartwood varies from rich red to reddish brown, darkening on exposure to light and producing an exceptionally smooth finish. Cherry is used in fine furniture, high end kitchen cabinets, paneling, boat interiors, and musical instruments.

Cottonwood - *(Populus deltoids)*

Cottonwood has characteristics similar to its Aspen cousin which, when turned on a lathe, will create a fuzzy surface requiring additional care when finishing. This wood is soft, weak when bent, and low in shock resistance. Cottonwood is used in venetian blinds, shutters, toys, and kitchen utensils.

Red Elm - *(Ulmus rubra)*

Red Elm has a grayish white- to light brown-colored sapwood; reddish-brown to dark brown heartwood, straight grain coarse texture, moderately heavy , hard, and excellent bending and shock resistant characteristics. Its main use is in furniture and cabinet making.

Gum - *(liquidamber styraciflua)*

The wood of this tree has irregular grain, usually interlocked, thereby producing an attractive pattern. Although easy to work with by both hand and machine tools, gum dries rapidly, has a strong tendency to warp and twist, and is susceptible to movement in performance.

Hackberry - *(Celtis occidentalis)*

A member of the elm family, hackberry is irregularly grained, sometimes straight-grained, with a fine uniform texture. Due to its fairly high shrinkage rate it is most suitable to be cut into small, short pieces of stock

Hickory and Pecan - *(Carya spp)*

This species is split into two groups-true hickories and pecan (fruit-bearing) hickories. The heaviest of American hardwoods, hickory is difficult to machine and glue and very hard for hand tool work. This wood has a tendency to split when nailed or screwed, so pre-drilling is mandatory. Hickory can be difficult to dry, has high shrinkage, and finds application in tool handles, ladders, dowels, sporting goods, flooring, and cabinetry.

Hard Maple - *(Acer saccharum, Acer nigrum)*

Also known as sugar maple, this wood is usually straight-grained, but also occurs with "fiddleback and "bird-eye" figures. Hard and heavy, maple machines and turns well. Hard maple is used in flooring, butcher blocks, worktops, kitchen cabinets, and tabletops.

Soft Maple - *(Acer rubrum, Acer saccharinum)*

Similar to hard maple, also known as red maple or silver maple, this species is about 25% less hard than hard maple and is often substituted for hard maple, stained to resemble cherry. It is also a substitute for beech.

Red Oak - *(Quercus spp)*

By far the largest species growing in the Eastern hardwood forests, red oak is more abundant than white oak. The sapwood of red oak is white to light brown and the heartwood is a pink reddish brown. Furniture, flooring, kitchen cabinets, door veneers, molding, and architectural millwork are made of this wood, the most widely used of the hardwood species.

White Oak - *(Quercus spp)*

Since this wood reacts with iron, galvanized nails are recommended for fastening. Otherwise similar in characteristic to red oak, white oak is less abundant.

Poplar - *(Liriodendron tulipifera)*

Creamy white sapwood and yellowish brown to olive green heartwood, this wood will tend to darken when exposed to light. A straight-grained wood that is easy to plane, machine, and turn, poplar takes stain and paint very well. Used in edge-glued panels, turnings, and carvings, this wood is also used in cabinetry and light construction.

Sassafras - *(Sassafras albidum)*

This wood is well known for its aroma. It has a coarse texture, generally straight-grained, and requires care in drying since it has a tendency to check with small movement in performance. Sassafras is rather scarce and is used in some furniture and millwork manufacture.

Sycamore - *(Platanus occidentalis)*

Also known as buttonwood, it machines well but high speed cutters are required to avoid chipping. This wood has a fine close texture with interlocking grain and finds use as drawer sides and other furniture parts, paneling and moldings, toys, and kitchenware.

Black Walnut - *(Juglans nigra)*

The sapwood of this species is creamy white while the heartwood is the more recognizable light brown to dark chocolate brown.

Some boards have wavy or curly grain but most are straight-grained. It holds stain well and presents an exceptional finish. Black walnut is sometimes used as an accent or contrasting wood when placed with lighter colored surfaces.

Willow - *(Salix spp)*
This wood works easily with both hand and machine tools, but presents a fuzzy surface when interlocked grain is present. The grain, while generally straight, often presents a "display" figure. Willow is weak in bending and shock resistance, but is a good substitute for walnut.

MILLWORK-DOORS AND FRAMES, WINDOWS, STAIRS, CABINETS, COUNTERTOPS

Millwork items have their own unique terminology and familiarity with the terms for each major component of millwork is not only interesting, but, on occasion, very helpful. There may come a time when you notice a particularly interesting detail on a door, frame, window, or cabinet and would like to describe it to a friend or someone at a home decorating shop or your contractor. Rather than say, "I saw some wood paneling installed on a wall that was about so high a wall and it had the nicest cap on it something like a chair rail. Do you have anything like that I can see?"

You would get a more rapid response if you said, "Show me some wainscoting and I'd also like to see if you have a matching wainscot cap and apron."

The following diagrams will be helpful in that regard:

Figure 22.4 General Millwork Profiles and Nomenclature

Figure 22.5 Anatomy of a Traditional Staircase

Figure 22.6 Conventional Stair Profiles and Parts

Figure 22.7 Panel Door Terminology and Determining the "Handing" of a Door

Figure 22.8 Door frame profiles (configurations) and Window frame profiles

Figure 22.9 More window profiles including the sash. This diagram shows a sash (Section B,E,C) that opens and closes and fixed glass light in the other openings.

Figure 22.10 Window stiles (vertical pieces) and rails (horizontal pieces) and muntin bar details

Figure 22.11 Wood window blinds and screen construction details

(text continues on page 425)

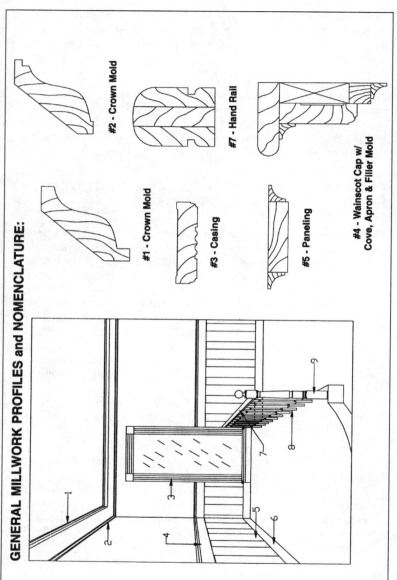

GENERAL MILLWORK PROFILES and NOMENCLATURE:

#2 - Crown Mold

#7 - Hand Rail

#1 - Crown Mold

#3 - Casing

#5 - Paneling

#4 - Wainscot Cap w/
Cove, Apron & Filler Mold

FIGURE 22.4 General millwork profiles and nomenclature. *(By permission-Woodwork Institute, West Sacramento, Ca., continued on next page)*

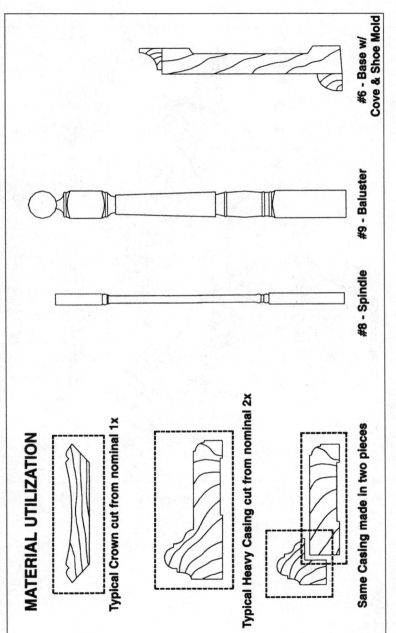

FIGURE 22.4 General millwork profiles and nomenclature *(continued)*.

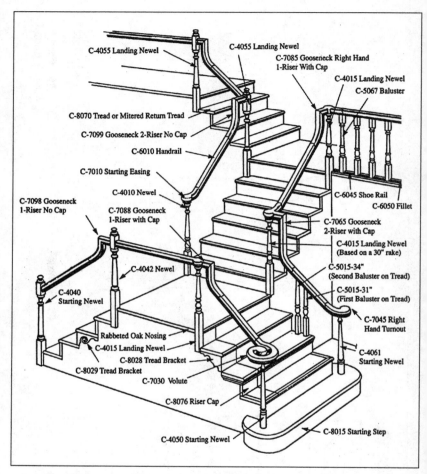

FIGURE 22.5 Anatomy of a traditional staircase.

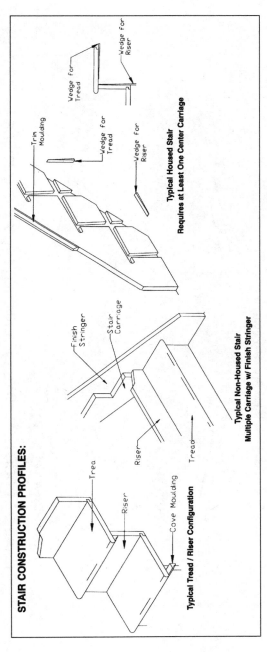

FIGURE 22.6 Conventional stair profiles and parts. *(By permission-Woodwork Institute, West Sacramento, Ca.)*

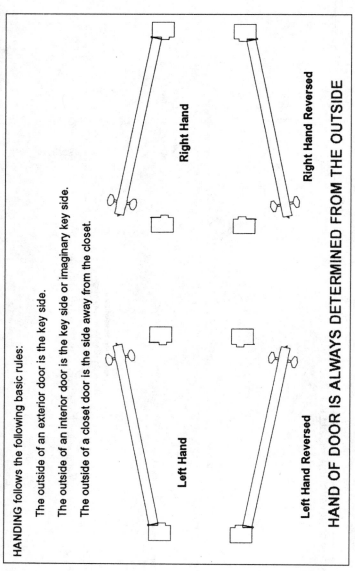

HANDING follows the following basic rules:

The outside of an exterior door is the key side.

The outside of an interior door is the key side or imaginary key side.

The outside of a closet door is the side away from the closet.

HAND OF DOOR IS ALWAYS DETERMINED FROM THE OUTSIDE

Right Hand

Right Hand Reversed

Left Hand

Left Hand Reversed

FIGURE 22.7 Panel door terminology and determining the "handing" of a door. (*By permission-Woodwork Institute, West Sacramento, Ca., continued on next page*)

HARDWOOD and SOFTWOOD doors shall be of special design and construction.

PANEL DOORS consist of stiles, rails and one or more panels.

GLAZED OR FRENCH DOORS consist of stiles, rails and one or more lights but may also contain one or more panels.

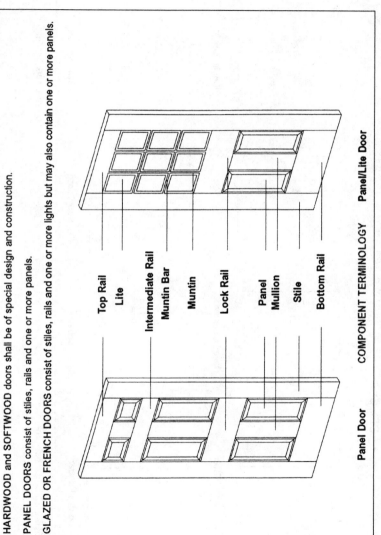

Top Rail
Lite
Intermediate Rail
Muntin Bar
Muntin
Lock Rail
Panel
Mullion
Stile
Bottom Rail

Panel Door   COMPONENT TERMINOLOGY   Panel/Lite Door

FIGURE 22.7 Panel door terminology and determining the "handing" of a door. *(continued)*

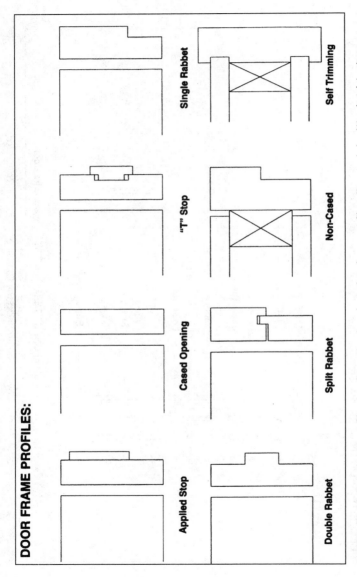

FIGURE 22.8 Door frame profiles (configurations) and window frame profiles. *(By permission-Woodwork Institute, West Sacramento, Ca., continued on next page)*

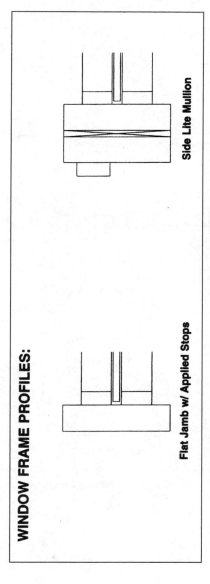

FIGURE 22.8 Door frame profiles (configurations) and window frame profiles. (*continued*)

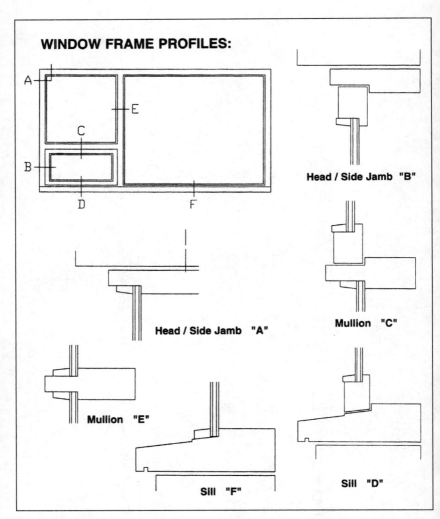

FIGURE 22.9 More window profiles including the sash. *(By permission-Woodwork Institute, West Sacramento, Ca.)*

TYPICAL WINDOW & SASH PROFILES

THE FOLLOWING PROFILES ARE EXEMPLARY OF GENERAL PRACTICE WITHIN THE INDUSTRY, AND ARE NOT INTENDED TO BE RESTRICTIVE NOR ALL-INCLUSIVE. Exact profile and species of lumber may vary to conform with local practice.

STOCK ITEMS, as produced by individual manufacturers, DO NOT NECESSARILY CONFORM TO THE STANDARDS FOR ARCHITECTURAL MILLWORK as established in this manual.

STILES and RAIL

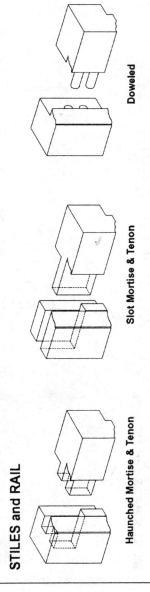

Haunched Mortise & Tenon

Slot Mortise & Tenon

Doweled

FIGURE 22.10 Window stiles (not styles), rails, and munton bar details sash. *(By permission-Woodwork Institute, West Sacramento, Ca., continued on next page)*

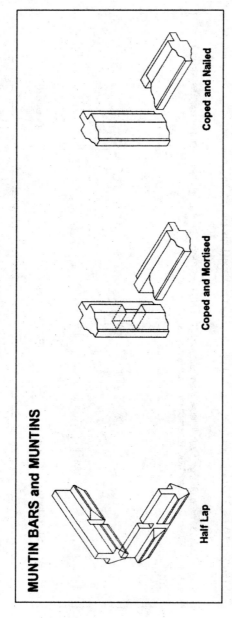

FIGURE 22.10   Window stiles (not styles), rails, and munton bar details sash. *(continued)*

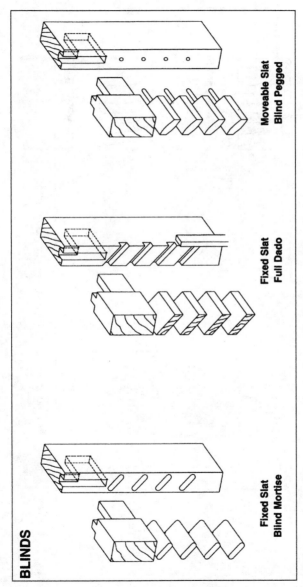

BLINDS

Fixed Slat
Blind Mortise

Fixed Slat
Full Dado

Moveable Slat
Blind Pegged

FIGURE 22.11 Wood window blinds and screen construction details sash. *(By permission-Woodwork Institute, West Sacramento, Ca., continued on next page)*

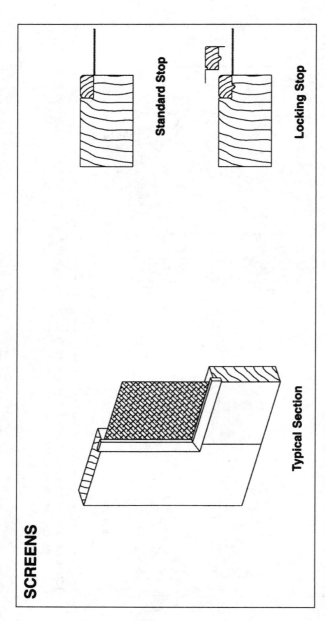

FIGURE 22.11 Wood window blinds and screen construction details sash. *(continued)*

Plastic Laminates

Not all millwork items are solid wood or wood veneer. Plastic laminates offer an expansive array of colors, patterns, and textures that simulate metals-both industrial and precious, stone, seashells, and wood patterns.

These finishes are relatively maintenance-free and, with proper care, require only an occasional wiping down with a household cleaner to preserve their appearance.

There are high pressure and low pressure laminates and several types of thermosetting materials, either applied in the factory or on the jobsite, to any of several substrates- wood plywood, particleboard, or medium-density fiberboard.

High Pressure Laminates

There are different grades of high pressure decorative laminates (HPDL) from recognizable manufacturers as Formica® Corporation and Wilsonart® and other such suppliers.

High pressure laminates can be either applied in the factory, or produced and sold as a finished product, i.e. countertops, produced in a millwork shop, or even fabricated by a carpenter on the jobsite.

These HPL (high pressure laminate) or HPDL (high pressure decorative laminate) products are available in three different grades:

- General purpose - most frequent used grade on countertops, work surfaces, front panels on workstations.
- Vertical grade - the grade of choice for wall cabinet fronts, door and drawer panels .This grade is suitable for vertical applications where a durable surface is not subjected to the same degree of wear-and-tear as required for horizontal surfaces.
- Postforming grade - for forming rolled edges of counter tops, backsplashes, cabinet doors, and drawer fronts where it is desirable to roll the laminate on a simple radius over the edge of the substrate.

These materials are available in sheets ranging from 36" to 60" (.9 to 1.5 meters) in width and 8 to 12 feet (2.4 to 3.6 meters) in length.

Low Pressure Laminates

Melamine, a low pressure laminate, is a thermosetting compound that is applied to a substrate like particleboard or medium density fiberboard (MDF). Unlike the laminate products above that are sold to the fabricator in sheets to be applied to substrates, melamine-surfaced panels are sold as panels and are cut to size as required and end banded where raw (uncoated edges) appear after being cut to size.

Polyester

This finish is actually a sprayed-on material usually applied over an MDF (medium density fiberboard) substrate. Polyester finishes are considered the most durable of all applied gloss finishes, but surfaces with this type application, when scratched or dented, can't be easily repaired without re-applying a new coating over the repaired area and possibly over the entire area to insure an acceptable match with adjacent panels.

Thermafoil

This is a thermally-activated vinyl material that is, in effect, shrink wrapped around a substrate, conforming to the configuration of that substrate. Thermafoil materials are available in white or a variety of wood look-alike veneers.

Cabinetry

Cabinetry and cabinet work refers to, as the name implies, the production of cabinets, whether free standing or built-ins. It also refers to the way in which pieces of wood in the manufacturing process are put together (joinery), whether it be wood-to-wood in the case of dovetailing or with metal fasteners-nails and screws. Figure 22.12 illustrates some common joinery methods.

Adhesives, you can't really call them glue anymore, are used in all types of joinery, frequently in conjunction with nails or screws. Figure 22.13 contains a list of various types of woodworking adhesives, uses, and characteristics.

When we think of cabinets, *kitchen cabinets* and *bath cabinets* come to mind. There are two terms that often arise when considering the purchase or manufacture of these types of cabinets - Frame and Frameless. What do these terms really mean?

Framed construction:
This type is easily recognizable if the doors to the cabinet are attached to a frame rather than the inside of the cabinet wall.

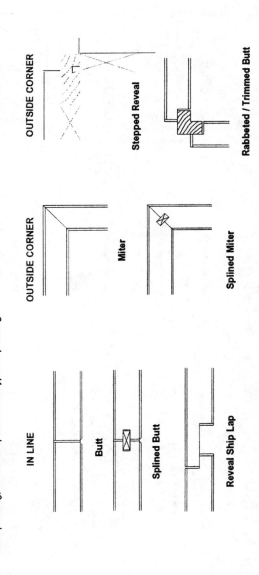

FIGURE 22.12 Joinery details. *(By permission-Woodwork Institute, West Sacramento, Ca., continued on next page)*

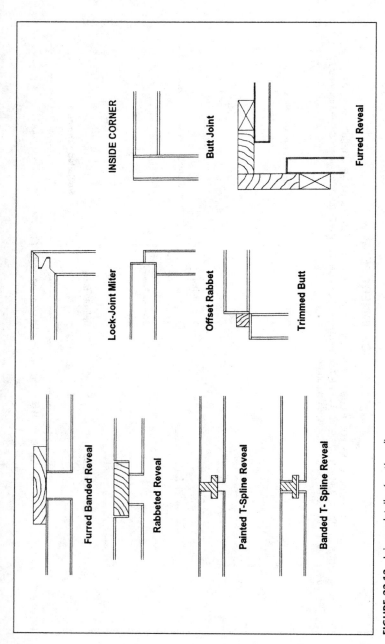

FIGURE 22.12 Joinery details. *(continued)*

ADHESIVE:			
PERFORMANCE TEST:	Type I-Fully Waterproof (Exterior)	2 Cycle Boil/Shear Test	
	Type II-Water Resistant (Interior)	3 Cycle Soak Test	
	Type III-Water Resistant (Interior)	2 Cycle Soak Test	

GENERIC NAME	USED FOR BONDING	ANSI/HPVA NWWDA -I.S.	CHARACTERISTICS
1. Aliphatic (Carpenter's Glue)	Wood and wood products	Type II	Non-toxic; non-flammable; and non-staining; waterproof; water resistant.
2. Casein	Wood and wood products	Type II	Highly water resistant, **NOT** waterproof.
3. Contact Cement	Plastic laminates and veneers to wood	Type II	Highly water resistant; **NOT** waterproof.
4. Epoxy	Wide range; wood; wood to metals	Type I	Two-part glue-formulas vary; Completely waterproof.
5. Hot-melt Glue	Wide range; bonds wood to vinyl, metal and wood	Not tested for moisture resistance	Liquefies when heated; bonds in a liquid state; solidifies as it cools. Used extensively for edge banders and other automatic equipment.

FIGURE 22.13 Guide to adhesives. *(By permission-Woodwork Institute, West Sacramento, Ca., continued on next page)*

6. Polyvinyl Acetate PVA	Wood and wood products	Slight moisture resistance	Good for cabinet work and interior woodwork. **NOT** recommended for joints with sustained loads.
7. Polyvinyl Acetate PVA Catalyzed	Wood and wood products	Type I	Used for assembly gluing where exterior waterproof bonds are required.
8. Polyvinyl Chloride PVC	Wide variety of materials	Not tested for moisture resistance	Crystal clear, fast drying.
9. Resorcinol Resin	Wood, wood products and laminates	Type I	Fully waterproof; purple glue line; two parts; liquid resin and powdered catalyst. Pot life-3 hours.
10. Urea Resin	Wood and wood products	Type II	Plastic resin glue; mixed with water; excellent for cabinet work; must be clamped. Drying time 3 - 7 hours at 70° F.
11. Panel / Construction Adhesive	Metal to wood, particleboard, or plywood; also plastic surfaces	Type II	Plastic epoxy base; liquid state; dries fast; very difficult to remove. Can be used to permanently set adjustment screws in European type hinges.

FIGURE 22.13 Guide to adhesives. *(continued)*

Framed cabinets have a framework of vertical and horizontal wood members joined together with various types of wood joinery methods such as in-line, butt, splined, or rabetted construction. The side and rear panels are usually held together with glue and screw fasteners with corner blocks in strategic places for added reinforcement. The doors of a framed cabinet will overlap the frame and generally be fastened with exposed hinges. A two-door cabinet will have a vertical wood stile inbetween the two doors. When all doors and drawers are removed, a framework is exposed, hence the name "framed construction." Figure 22.14 depicts a typical array of framed kitchen cabinets.

This type cabinet is frequently referred to as a "European Style" and does not have a frame. The doors are hung directly onto, or more accurately *into,* the box, usually on the inside face of the side panel. When these type of cabinets have movable shelves, they are secured by pushing stops into one of a series of vertically drilled

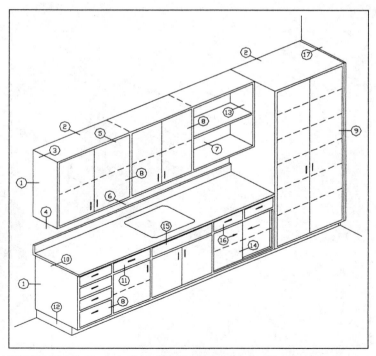

FIGURE 22.14 Typical cabinet layout. *(By permission-Woodwork Institute, West Sacramento, Ca.)*

holes on the inside of the side panel which permits easy up and down shelf adjustment. Because the doors are hung from hinges attached to the inside of the cabinet, this style cabinet has a more contemporary design, even though some door fronts may have a colonial or traditional motif.

Frameless construction:

The hardware on frameless cabinet doors is adjustable in two planes, in and out and up and down, and the close tolerances and adjustments required during installation make these cabinets a little more difficult to install, but the actual cost of installed frameless cabinets is less expensive than framed ones.

Figure 22.15 depicts the same array of kitchen cabinets shown in Figure 22.14, but in "frameless" form.

The Term "Exposed to View"

Some cabinet manufacturers produce cabinets with two degrees of finish, referred to as finish of areas exposed to view, semi-con-

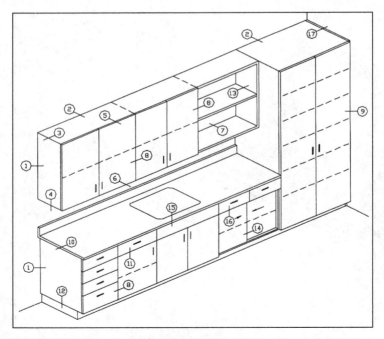

FIGURE 22.15 An array of frameless kitchen cabinets. *(By permission-Woodwork Institute, West Sacramento, Ca.)*

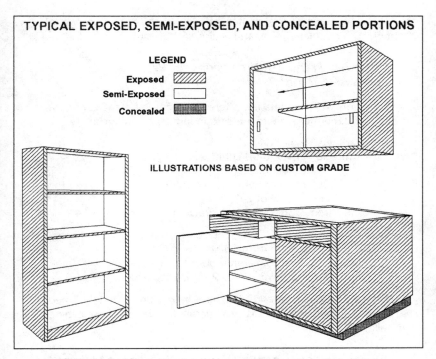

FIGURE 22.16 Exposed to view or concealed from view. *(By permission-Woodwork Institute, West Sacramento, Ca.)*

cealed, or fully concealed. When exotic or expensive finishes are required for cabinets, some areas are not exposed to view, such as the back panel of a cabinet with doors, or the read surface of a back panel that will be fastened to the wall. Manufacturers often use less expensive materials for millwork parts that are "concealed" from view or semi exposed. In Figure 22.16 the cross hatched areas are considered exposed to view; the solid shaded areas are concealed and can therefore be of lesser quality (appearance-wise) as can some of the semi-exposed areas.

Countertops

Kitchen cabinet counter tops can be constructed of a variety of materials ranging from wood butcher block to high density laminates -HDL- (high density laminate-Formica® or Wilsonart®, for example) to ceramic tile to granite or other types of stone.

HDL- high density laminates offering such a wide range of options is a popular choice, and is also cost effective.

What Are Standard Dimensions for Wall and Base Cabinets?

Although wall cabinets are offered in many different configurations and dimensions, there are certain industry standards:

Base cabinets:
Height - 36" at stand-up counters
31" for vanities
30" for sit-down counters with a
knee space of 24½"
Depth - from front of cabinet drawer to face of wall - 24"

Wall hung cabinets - Height - including light apron, if required, 36", but can range from 18" to 36"
Depth - from front of cabinet door to face of wall - 24" standard, but depths of 12" to 24" also available.

Figure 22.17 illustrates the minimum masonry, metal-framed, and wood-framed wall construction details for wall and base cabinet installation.

Finishing Cabinetry and Wood/Wood Veneer Millwork

The beauty of wood is preserved and enhanced when proper finishing techniques are employed. Millwork manufacturers generally offer four levels or grades of finish: economy, custom, premium, and laboratory.

Economy establishes the standard to meet the requirements of lower cost residential and commercial construction.

Custom includes all of the procedures to produce a high quality product suitable for use in higher quality residential, school, and commercial work.

Premium finishing is a superior quality of workmanship and materials, with a price to match.

Laboratory, not generally used in residential work, presents a chemical resistant finish.

Open Grain and Close Grain Finishing Techniques

One of the classifications for hardwoods is open or closed grained and finishing techniques for each type varying slightly depending upon this classification.

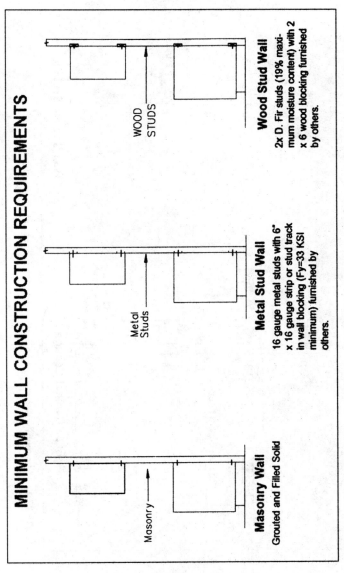

FIGURE 22.17 Minimum wall construction and cabinet attachment details. *(By permission-Woodwork Institute, West Sacramento, Ca., continued on next page)*

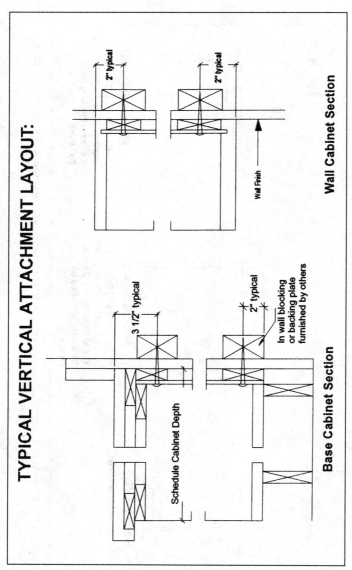

FIGURE 22.17 Minimum wall construction and cabinet attachment details. *(continued)*

Open Grain - ash, butternut, Chestnut, African mahogany, White and Red Oak, Walnut and Honduras mahogany

Close Grain - red alder, beech, birch, cherry, gum, maple

Open grain materials require a filler prior to the application of a smooth coating; close grain boards and most softwoods do not require a filler. Many finishing materials require additional steps when applied to either open grain or closed grain finishes - so read the label on the container carefully to determine whether any such added steps are necessary for your work.

Types of stains:

- Lacquer stains - recommended for Economy grade finish only. Blotching and uneven color and grain pattern uniformity will occur. The lighter the stain, the better the end result.

- Dye stains - used on premium hardwood (solid wood, not veneer). It offers no blending capabilities but will result in the clearest grain color and pattern. It can be toned after the sealer coat. Note: This type stain should only be applied by a professional.

- Oil based stains - the best stain for color uniformity, it can be wiped or brush-blended. When there is a difference in wood color and type, either solid or veneer, brush blending of the stain will create overall uniformity.

- Paste wood fillers - This material should only be used by skilled applicators

- Non-grain raising stains - contain no pigmented solids and are usually spray applied

- Wiping stains - contain color particles and may be applied by spraying, hand wiping, or brushing

fastfacts

Teak does not accept finishes in the same manner as other hardwoods; the most common finish is with penetrating oil without any filling or sealing.

fastfacts

Dark stains on birch and maple are not recommended, but if desired, the wood should be filled and/or wash coat-sealed before applying stain.

Some Finishing Terms

Bleaching - a process that lightens the base color of the wood to give it a uniform appearance

Fillers - used to close or fill pores to give the wood a smooth appearance. Applied by brush, roller, or spray or wiped off against the grain

Glazing - a specialty process to achieve color uniformity when the natural wood may be too strong in contrast

Toning - the use of a semi-transparent color to block out or reduce the color of the wood

Sealers - a finishing material applied to stop the absorption of succeeding coats; locks in the stain and also provides a base for the final top coat

Wash coats - thinned coats of sealer acting as a barrier against over penetration of stains, a condition that can cause blotchiness.

Iron stain - a dark bluish to black color caused by the natural tannic acid in the wood, primarily oak, when it comes in contact with iron or moisture. To prevent iron stain, never use steel wool on bare wood and, if shellac is used, it should not have been stored in an iron container. To remove iron stain, apply a solution of 12 ounces of Oxalic acid crystals dissolved in one gallon of lukewarm water. Apply with rubber gloves and goggles. Allow to dry and then sand with 150-180 grit sandpaper; thoroughly rinse with water and allow to dry.

Hand rubbing - provides a smooth, uniform finish, but adds cost to the finishing process.

High polished finish - involves several operations of wet sanding, buffing, and high gloss polishing at considerable cost.

INDEX